Multiscale Optimization and Materials Design

Frontier Research in Computation and Mechanics of Materials and Biology

ISSN: 2315-4713

Published:

Vol. 3 *Multiscale Optimization and Materials Design*
by Jun Yan & Gengdong Cheng

Vol. 2 *Advanced Continuum Theories and Finite Element Analyses*
by James D Lee & Jiaoyan Li

Vol. 1 *Introduction to Practical Peridynamics: Computational Solid Mechanics Without Stress and Strain*
by Walter Herbert Gerstle

Frontier Research in Computation and Mechanics of Materials and Biology – Vol. 3

Multiscale Optimization and Materials Design

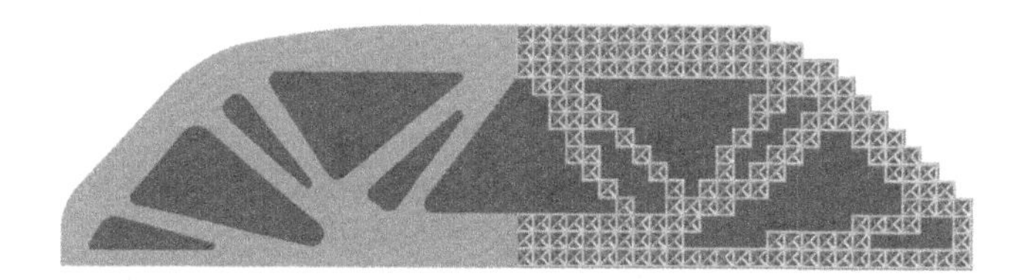

Jun Yan

Gengdong Cheng

Dalian University of Technology, China

NEW JERSEY • LONDON • SINGAPORE • BEIJING • SHANGHAI • HONG KONG • TAIPEI • CHENNAI • TOKYO

Published by

World Scientific Publishing Co. Pte. Ltd.
5 Toh Tuck Link, Singapore 596224
USA office: 27 Warren Street, Suite 401-402, Hackensack, NJ 07601
UK office: 57 Shelton Street, Covent Garden, London WC2H 9HE

British Library Cataloguing-in-Publication Data
A catalogue record for this book is available from the British Library.

Frontier Research in Computation and Mechanics of Materials and Biology — Vol. 3
MULTISCALE OPTIMIZATION AND MATERIALS DESIGN

ISBN 978-981-121-653-4 (hardcover)
ISBN 978-981-121-654-1 (ebook for institutions)
ISBN 978-981-121-655-8 (ebook for individuals)

For any available supplementary material, please visit
https://www.worldscientific.com/worldscibooks/10.1142/11721#t=suppl

Printed in Singapore

Dedication

The research in this monograph has been carried out for more than ten years, and it was originally from the National Natural Science Foundation of China's key project (10332010) on the new methodology of the ultralight materials and structural design. Then, I subsequently receive continuous funding from the NSFC, including (90816025, 10902018, 11372060, 11672057, 11711530018, U91906233). These funds provide key funding and direction for research contents. At the same time, with the continuous deepening of basic research, the corresponding achievements have received extensive attention from the industry, and which further provides me with funding and application supports. These cooperations are very appreciated, and they are included China Aviation Development Commercial Aviation Engine Co., Ltd., China Space Technology Research Institute, and China National Offshore Oil Corporation. With the supports of these cooperative institutions, I can't only carry out continuous and systematic research in the field of basic scientific research but also be able to obtain preliminary engineering applications. It is undoubtedly very pleasure to be able to see your design become a real product and be applied.

My research benefited from the excellent research atmosphere of the Department of Engineering Mechanics and the State Key Laboratory of Structural Analysis of Industrial Equipment of the Dalian University of Technology. I'm also very grateful for the many discussions and valuable feedback from my colleagues, Profs. Bo Wang, Xu Guo, Shutian Liu, Ling Liu, Bin Niu, Gang Li, Dixiong Yang, etc. The frank discussions and heated arguments at the group meeting every weekend helped to improve my research work. It's actually your support and encouragement that has promoted the writing of this monograph.

I would like to thank Prof. Shaofan Li of the University of California, Berkeley, who gave me very good suggestions when he learned of my research work, and encouraged me to publish the research work as a series of monograph, so that more people could systematically understand our research.

I'm very much grateful to Prof. Gengdong Cheng, who is the co-author of this monograph. While we together formulate the PAMP framework for concurrent multiscale topology optimization of structure composed of structured material, as the supervisor of my Ph.D., he guided the direction of this research, encouraged me to constantly and deliberately innovate in the research and to pursue perfection, especially to give me support when the research was in trouble.

I would also like to thank Mr. Steven Patt of the publisher for his hard work, discussing with me the plan for publishing the monograph, and removing obstacles and difficulties in the process.

Finally, I'm indebted to the students in my laboratory, who are Ph.D. students Zunyi Duan, Qianqian Sui, Zhirui Fan, Qi Xu and master's students Wenbo Hu, Suxia Yang, Jingyuan Wang, Jingzhao Liu, Sixu Huo, Quanyi Yin, Cuncun Jiang, Zhihui Liu, Fuhao Wang and those who pointed out the errors in this book.

Jun Yan
September 24, 2020

Contents

List of Tables

List of Figures

Chapter 1

Introduction

1.1 *Background and motivations*

Lightweight structures can reduce the long-run energy consumption of vehicles such as aeronautic crafts, automobiles, and ships. With the shortage of energy, resources, and intense competitions, lightweight structures have attracted more and more attention in aerospace, automotive, and ocean engineering. For high-speed vehicles, structural weight is one of the decisive factors for their essential performances, such as speed, efficiency, and maneuverability. In general, there are two main approaches to realize the structural lightweight design, which are the structural optimization design and the selection of lightweight materials. Structural optimization design has undergone tremendous development in the past decades, not only in theory and method but also in engineering applications. Especially, structural topology optimization [1, 2] is a powerful approach to realize an innovative lightweight structural design. Another approach of the structural lightweight design is the selection of new lightweight materials, such as Al-alloy, Al-Mg alloy, Ti-Alloy, and composite materials.

With the rapid development in manufacturing technologies, many lightweight materials with internal microstructures have been increasingly utilized in wide application areas as load-bearing members, heat exchangers, energy absorbers, and key components of aircraft engines, etc. An important kind of such material is the so-called "structured materials", which features constructions with periodical microstructures (e.g., metallic honeycomb, truss-like lattice materials, cellular alloys, and glass or carbon fiber reinforced polymers (GFRP/CFRP) laminates). In addition to multifunctionality, structured materials also demonstrate the unique designability both the microscale (i.e., the design of microstructural patterns) and the macro-scale (i.e., the design of structural configurations).

As Ashby [3] pointed out, 'The more efficient design is taking into account the macro-structure and micro-material selection at the same time'. The emergence of various structured materials provides a vast opportunity for integrated design of structures and their components, and the interaction between the designed structures and materials can be considered. It can be expected that a global optimum design, or an ideal lightweight structure, should be composed of the different materials with consideration of its various geometrical size, shape, and boundary/loading conditions, and vice versa. That is the size, shape, and even topology of the optimial structure needs to be changed correspondingly when its composed material is different. As inspired by nature, it has been found that the material of natural biological structures such as animal's bones and plant's stalks has a point-wise varying microstructure [4], which is a typical integrated design of the structured materials. As such, there is an utmost need for integrated optimization methodologies to achieve an optimal macro-structure with an optimal microstructure given any specific design objectives and constraints.

Multiscale design optimizations of the macro-structure and microstructure of the structured materials for global and local structural performances are studied in this book as a representative methodology to achieve the integrated design of structures and materials. Three critical challenges that are unique to the multiscale nature of the proposed optimization framework need to be addressed in the established design approach. First, the number of dimensions of the structural design space multiplied by that of the microstructural design space leads to a huge amount of dimensions for the multiscale design, making it almost unsolvable by conventional single-scale design approaches. Second, even with advanced manufacturing technologies such as 3D printing, manufacturability is still a topic of concern, making it necessary to have some control over the microstructural optimization in the multiscale design. Third, due to the multifunctional utilization of lightweight structural materials, it is essential to consider mechanical, dynamical, and thermal objectives and constraints, and their combined effects on the multiscale optimization.

A computationally effective solution framework for the multiscale structural optimization, especially topology optimization, had been established to allow a concurrent design of configurations of the macro-structure and the material's microstructure. To address the fundamental challenges arising from the ultra-large dimensions of the design space, this work proposed a novel Porous Anisotropic Material with Penalty (PAMP) approach to decouple the microscale design from the macro-scale design, which is then integrated into a unified multiscale optimization framework through a mathematical homogenization. An efficient optimal search was enabled by the PAPM model along with analytical multiscale sensitivity analysis and corresponding numerical schemes. Through the innovative topology optimization design frameworks, some novel microstructural configurations with anisotropic features can be discovered along with their optimal distributions at the macro-scale for optimal performances in the target applications.

The PAMP approach as the pioneering framework on multiscale design optimization has been extensively utilized to solve different kinds of multiscale topology optimization problems due to its effectiveness, efficiency, and robustness: (1) Optimization of stress constraints; (2) Optimization of dynamic performances; (3) Optimization of composite structures; and (4) Optimization of thermoelastic structures.

It is worth to point out that manufacturing factors are strongly underlined in this book by assuming homogeneity of the material microstructures at the macro scale to improve the possibility of practical applications (except the Chapter 6: Composite frame structure optimization).

1.2 *Literature review*

1.2.1 *Multiscale topology optimization of structured materials with structural stiffness, strength and dynamic objectives*

Since the structures composed of structured materials usually include a large number of micro-components, the structural modeling and analysis with conventional finite element techniques are not applicable, especially considering the optimization of the structure [5, 6] and materials. It is necessary to develop new numerical simulation methods and optimization design frameworks for engineering structures of structured material. An equivalent analysis based on the multiscale method is needed. For an equivalent analysis, homogenization, multiscale finite element method (MsFEM), and representative volume element (RVE) methods have been widely used. State-of-the-art researches for homogenization, MsFEM, and RVE approaches can be found in Hassnai et al. [7], Bendsoe et al. [8], Guedes et al. [9], Haftka [10], Huet [11], Balendran et al. [12], Hou et al. [13].

Asymptotic Homogenization (AH) method [7] is a perturbation-based mathematical approach to calculate the equivalent mechanical properties of structured material numerically. However, the AH formulation is not easy to combine with FEM and take the advantage of available commercial software of FEM [14]. The Novel Implementation of Asymptotic Homogenization (NIAH) method proposed by Cheng et al. [15] improves the computer implementation of traditional AH approach in structural optimization [16]. It replaces the imposed eigenstrain by the corresponding nodal displacements and calculates the equivalent elastic properties by using the nodal characteristic force and displacement, which can easily be extracted from FEM calculation.

Compared with the AH method [17, 18] in which the size of material microstructures was assumed to be infinitely small, the MsFEM developed by Zhang, Hou et al. [6, 13, 19] was applied to the structural analysis of structured materials, periodic lattice materials successfully. The main idea of MsFEM is to construct finite elemental shape functions numerically that can capture the microscale heterogeneity of materials to achieve macro-scale solutions. The effects of the finite size of the material microstructure can be considered in the MsFEM. Meanwhile, the downscaling calculation of stress at the micro-component is relatively easily implemented.

Base on the multiscale structural analysis approaches, the topology optimization can be implemented on the macro-structural and microstructural scales concurrently to find the optimal configurations for the microstructures of the structured materials and the optimized distribution of the materials on the macro-structural domain. The pioneering work of continuum topology optimization can be traced back to Cheng and Olhoff's research [1] about optimal distribution of thickness for minimum compliance of an elastic plate. Bendsøe and Kikuchi [2] initiated topology optimization based on the AH theory. Then the AH method [7, 20] and density-based methods, such as solid isotropic material with penalization (SIMP) [8, 21] have become two popular methods in the field of topology optimization. The above studies focused mostly on structural optimization composed of solid isotropic materials. However, with the fast development of manufacturing techniques [22] as mentioned above, ultralight structured materials are widely manufactured and applied in the industrial structures [23].

The optimization of a structure composed of the structured materials is an interactive coordination process to meet the increasingly strict demands for the lightweight design of vehicle structures. Rodrigues et al. [24] obtained the optimal hierarchical designs of structures composed of structured materials via a AH method on both structural and material scales. Coelho et al. [25] expanded the optimal hierarchical scheme to 3D structures. However, the optimized microstructural configurations may vary considerably from point to point in macro-scale. It will result in diverse and disconnected material microstructure, which increases the manufacturing difficulty. Liu, Yan and Cheng [26] proposed the PAMP (Porous Anitsotropic Material with Penallty) approach and optimized structures composed of porous materials with a concurrent topology optimization method for minimum structural compliance assuming the uniformity of microstructure in macro-scale. A similar optimization model was applied by Niu et al. [27] to maximize the fundamental natural frequency of 2D structures composed of structured porous materials. Pizzolato et al. [28] presented a topology optimization method to design assemblies of periodic cellular materials with controllable geometric complexity, in which the structure, the layout of the material subdomains, and their micro-structures are optimized concurrently. Zhao et al. [29] presented an efficient concurrent topology optimization approach based on PAMP for minimizing the maximum dynamic response of two-scale hierarchical structures in the time domain. And an enhanced decoupled sensitivity analysis method was proposed for concurrent topology optimization of the time-domain dynamic response problems. A concurrent multiscale multi-material topology optimization method for minimizing sound radiation power of the vibrating structure subjected to harmonic loading was presented by Liang et al. [30]. In order to improve the structural performance, Gao et al. [31] used a parametric level set method to optimize the topologies of the macro-structure and material microstructures, with the effective macro-scopic properties evaluated by the AH. Fu et al. [32] presented a multiscale level set topology optimization method based on numerical homogenization for designing a shell-infill structure. Deng et al. [33] developed a robust concurrent TO (topology optimization) approach for designing multiscale structures composed of multiple porous materials under random field loading uncertainty. Cai et al. [34] proposed an efficient method for robust concurrent topology optimization of the multiscale structure under single or multiple load cases based on the bi-directional evolutionary structural optimization (BESO) method. However, the multiscale optimal designs of structures composed of lattice materials are rarely reported. Xia et al. [35] considered a truss-like structure as a framework of many short beams, and a triangular mesh is used as an initial ground structure for the optimization of 2D truss-like structures. Zhao et al. [36] developed a concurrent design method of Additive Manufacture (AM) fabricated lattice structures and its printing part orientation in natural frequency optimization problems. Meanwhile, it also should be pointed out that most of the above studies were based on a AH method in which the size of the material microstructure is infinitely small, and the size effects of the microstructure of the porous materials are omitted for the structural analysis and optimization. In recent years, some researches studied the size effects of the microstructure of the structured materials on the results of structural analysis and optimization [37-39]. Tekoglu and Onck [40] pointed out that the mechanical properties of a structure composed of porous materials are strongly dependent on the ratio of macro-structure to the size of the unit cell. Xie et al. [41] found that the optimized topologies of the unit cells converged rapidly to certain patterns by increasing the number of unit cells. Lipperman et al. [42] implemented the maximum strength design for a complex ground structure that is composed of a lattice material with actual size under periodic boundary conditions in the uniaxial stress field.

1.2.2 *Multiscale topology optimization of thermoelastic structures*

The design of thermoelastic structures is a basic and important issue in aerospace industries. Comparatively, topology optimization of thermoelastic structures is relatively more complicated because it belongs to a type of design dependent problem. That means the thermal load changes with the changing of the spatial distribution of solid material phases. Here the thermoelastic structure is loaded both mechanical and thermal loads simultaneously. Many studies have been devoted to the topology optimization design of thermoelastic structures to achieve the lightweight design and improve the structural performances [2, 43-45] simultaneously, such as the thermal stress, thermal deformation, and the frequency of the thermoelastic structure. Using the structural compliance as an objective function, Rodrigues and Fernandes [46], Pedersen et al. [47], Li et al. [48], Deaton et al. [49], Wang and Cheng [50] investigated the optimum design of cellular material distribution at maximization of heat dissipation rate while minimizing the prescribed flow pressure. Gao and Zhang [51] studied the optimization of structural performance under both mechanical and thermal loads. Xia [52] implemented a topology optimization of thermoelastic structures with the level set method. The objective is to minimize the mean compliance of a thermoelastic structure with a material volume constraint. Pedersen and Pedersen [47] studied the strength optimization design of a thermoelastic structure with the maximum Von Mises stresses as an objective. Zhang et al. [53] studied the differences between the minimum structural compliance and minimum structural strain energy during the topology optimization of thermoelastic structures from the aspect of sensitivity analysis and topology configurations. They found that the minimum structural strain energy was a better objective to reduce the structural stress for the thermoelastic structure. Deaton and Grandhi studied the minimum weight design of a thermoelastic continuum structure with stress constraints. Wang et al. [54] proposed a multi-objective optimization model that combines low thermal directional expansion with high structural stiffness. Gao et al. [55] studied on the topology optimization design problem in a structure composed of multiple materials under the conditions of steady-state temperature and mechanical loading. Wu et al. [56] addressed this issue by proposing multi-material topology optimization for thermo-mechanical buckling problems. With considered temperature variation effect, Zhu et al. [57] carried out the shape-preserving design successfully in thermoelastic problems to prevent local thermal damages.

However, in most of the above works, macro-scale structural optimization is emphasized, and the coupling effect between the structures and materials is not considered. In fact, because of excellent performances in thermal insulation and lightweight, some structured materials are manufactured and available in the market. They have almost the homogeneous microstructure and are widely applied in the equipment of electronic and aerospace industries etc. [41, 58, 59]. Many studies [60] showed that when thermal-mechanical effects are taken into account, structures composed of porous materials with well-designed microstructures are promising candidates to enhance structural/thermal performances.

By applying a three-phase topology optimization method, Sigmund and Torquato [61] achieved the microstructural design of materials with extreme thermal properties. Deng et al. [60] proposed a multi-objective optimization model for a thermoelastic structure composed of light porous materials to simultaneously improve a structural stiffness and reduce the thermal expansion of the structure in a particular direction. Radman et al. [62] introduced an alternative approach for the topological design of microstructures of materials that are composed of three or more constituent phases. It is assumed that the materials are made up of periodic microstructures. Pelanconi et al. [63] focused on the application of a ceramic tubular high-temperature heat exchanger with engineered cellular architectures. Niknam et al. [64] investigated the effect of cell architecture on the bending behavior of architected cellular beams subjected to a thermo-mechanical load. Wu et al. [65] established a multiscale topology optimization method for the optimal design of non-periodic, self-supporting cellular structures subjected to thermo-mechanical loads. The result is a hierarchically complex design that is thermally efficient, mechanically stable, and suitable for additive manufacturing. Xu et al. [66, 67] developed a concurrent topology optimization scheme based on BESO, which could be applied to composite thermoelastic macro structures and microstructures with multi-phase materials. Recently, the PAMP optimization framework has turned out to be an important method in much more applications, such as the design of graded lattice structure [68, 69].

1.2.3 *Multiscale design optimization of fiber-reinforced composites*

In recent years, fiber-reinforced polymers with glass or carbon fiber (GFRP/CFRP) have been widely used in engineering [70] as the representative of new lightweight materials. In engineering design practice, it is found that the development of multiscale optimization theories and methods for structures composed of composite materials is a very attractive and rewarding research area. Except the size, shape, and topology optimization for macro structures as described above, the microscopic parameters of the composites can also be designed, such as its fiber laying angle, thickness, and stacking sequence. This characteristics provide an extended design space for designers to improve the structural performances by adjusting the microscopic parameters of the composite materials.

The optimization theory and technique for microscopic parameters of composites have attracted much attention in the last decades [71-74]. Many researchers adopted genetic algorithms to optimize composite laminates [75-77]. The fiber orientations and fiber volume fractions of the laminate were chosen as the primary optimization variables. In addition to the intelligent algorithm, Graesser et al. [78] combined a state-of-the-art global optimization algorithm with classical lamination theory to optimize composite laminates. A state-of-the-art review and recent developments in this field can be found in the review articles of Ghiasi et al. [79, 80], Bakis et al. [81], Ganguli [82], Nikbakt et al. [83] and Xu et al. [84].

Unlike the traditional optimization of composite structures, which considers continuous uniform fiber distribution in one ply, a discrete fiber ply angle is considered as one of the manufacturing constraints. Stegmann and Lund [85] introduced an efficient method to realize a single choice of material among a finite set of candidate materials using Discrete Material Optimization (DMO). Sørensen et al. [86] presented simultaneous optimization of the thickness and fiber orientation of laminate composites. Based on discrete material and thickness optimization (DMTO), Wu et al. [87] realized the simultaneous design of ply orientation and thickness of laminated structures with adopting casting-based explicit parameterization to suppress the intermediate void across the thickness of the laminate. Niu et al. [88]. dealt with vibro-acoustic optimization of laminated composite plates with DMO method to minimize the total sound power radiated from the surface of the laminated plate to the surrounding acoustic medium. Kiyono et al. [89] proposed a normal distribution fiber optimization (NDFO) method to guarantee total fiber convergence in the selection of a candidate angle. Recently variable-stiffness design with continuing fiber path has attracted much attention as Hao et al. [90], Akhavan and Ribeiro [91], Hao et al. [92], Shafighfard et al. [93], Serhat et al. [94]. Bruyneel [95] proposed a new multi-material parameterization named SFP (Shape Functions with Penalization), where bi-linear finite element shape functions act as weights to interpolate among the candidate materials. Gao et al. [96] proposed another multi-material parameterization scheme named BCP (Binary Coded Parameterization), where any number of candidate materials can be considered. Recently, Hvejsel and Lund [97] noted the indistinct material selection problem, and they discussed and generalized the SIMP and RAMP schemes using a large number of sparse linear constraints to limit the material selection to at most one in each domain. Gao and Zhang [98] found that the normalized weights of the design variables weaken the punishment and make it difficult to converge to a clear fiber ply angle choice in discrete composite optimization. Recently, Hao et al. [92, 99, 100] had carried out a series of researches on complex variable-stiffness (VS) shells design based on the degenerated shell method using the isogeometric analysis method. The non-convergent problem has been recognized in previous references, and several methods were proposed, but they remain insufficient. The multi-materials interpolation schemes must be further studied and enriched to improve the convergence rate of the fiber angle selection and the convergence speed of the optimization iteration.

To gain the best match of the composite material and structure and realize an integrated optimization of a composite material parameter and its distribution in the macro-structural domain, following the basic idea of PAMP, the fiber ply angle in one element can be considered as one micro design variable, and the materials that exist or do not exist in the design domain are considered as the macro design variable in the content of multiscale optimization of composite structures. Considering the costs and the process requirements, the most commonly used discrete fiber ply angles $[0, \mp 45, 90]$ are chosen as the microscopic candidate materials to establish the optimization model of minimum structural compliance with a specified composite volume constraint. To overcome the difficulty of non-convergent elements in the traditional DMO approach and obtain a clear fiber ply angle choice, an improved discrete material penalty model, which is labeled HPDMO (Heaviside Penalization of Discrete Material Optimization), was proposed by Duan et al. [101]. The modified Heaviside penalty function [102-104] is introduced to replace the polynomial material interpolation formula of the material/structural optimization. The HPDMO model effectively overcomes the challenge using a discontinuous fiber ply angle based on the classical structural optimization algorithm with derivative information. The numerical examples show that compared with the DMO model, the improved HPDMO model can achieve the optimal result through fewer iteration steps and improve the convergence rate of the fiber angle to obtain clearer optimal results.

1.3 *Outline of the book*

Concurrent multiscale design optimization of a structure composed of structured materials is studied in the present book. The aim is to obtain the optimal configuration of the microstructure of the structured materials and its optimal distribution in the macro-structural domains for better structural performances. Global structural performance such as minimum compliance and maximum fundamental frequency design, local performance such as stress concentration, and multi-objective performances such as thermal deformations are investigated. This book is organized as follows:

In Chapter 2, a concurrent multiscale optimization framework PAMP for structures composed of structured materials is proposed. The topology optimization is implemented to simultaneously achieve the optimized configurations of macrostructures and material microstructures for minimizing structural compliance. In the optimization, AH method and EMsFEM are employed to connect the two geometrical scales. Several numerical exmaples are introduced to emphasize the advantages of the proposed PAMP framework.

In Chapter 3, optimizations of structural strength problem based on PAMP optimization framework is performed. Three different optimization objectives are firstly introduced to handle the stress minimization problem, where the lattice material is modeled by micropolar continuum representation by considering the possibility of high stress gradient in the optimization. Subsesquently, lightweight design under stress constraints based on EMsFEM is studied. The size effect of lattice structures is then discussed in detail and thereby some interesting findings are given.

In Chapter 4, the multiscale optimization for maximizing structural fundamental vibration frequency is formulated based on PAMP and AH. To eliminate the local vibration mode in the low-density area of the intermediate design the polynomial interpolation scheme is developed. The volume preserved Heaviside projection enables to obtain stable iteration while achieve black-white design. Several numerical examples for finding optimal configurations of macro structures and microstructures of cellular materials with the maximum structural fundamental frequency demonstrate the effectiveness of the method.

In Chapter 5, the single-objective and multi-objective optimizations of themoelastic structures are carried out and the merits of porous material is then revealed in-depth. In the single-objective optimization, the the effect of the temperature on optimized results in micro and macro scales is studied. Benchmark examples are given to reveal the rationality of the variation of optimized results as the temperature changes. Subsesquently, the advantages of the concurrent optimization model in improving the multi-objective performances of the thermoelastic structures are investigated. It is found that an “optimal” material volume fraction is observed in some cases when both the structural compliance and thermo deformation are considered in the objective function.

In Chapter 6, the concurrent multiscale optimization design for the minimum structural compliance and maximum fundamental frequency of fiber-reinforced frame structures are implemented. Due to manufacturing requirements, discrete fiber winding angles are specified for the micro design variable. The improved Heaviside penalization discrete material optimization (HPDMO) interpolation scheme has been applied to achieve the discrete optimization design of the fiber winding angle. Six types of manufacturing constraints are explicitly included in the optimization model as a series of linear inequality or equality constraints. Semi-analytical sensitivity analysis method is employed to accelerate the optimization process. The optimization results of the fiber winding angle and the macro-structural topology on the single scale, together with those of multiscale optimization, are studied and compared in the numerical examples with consideration of a composite plate, shell, and frame structure.

Chapter 2

Static: Stiffness Optimization

In this chapter PAMP (Porous Anisotropic Material with Penalty) optimization framework is proposed to formulate the concurrent multiscale stiffness topology optimization of structure composed on structured material and realize the multiscale structural and material designs. Two methods, AH method and EMsFEM are applied to find the effective material properties of structured material and thus establish the link between macro structural topology and microstructure topology of structured material.

In the optimization based on homogenization, the effective material properties and its sensitivity with respect to design variables are obtained by AH method. Several numerical examples are carried out to investegate the trade-off between the optimum macro and micro designs.

In the optimization based on EMsFEM, the effects of linear and periodic boundary conditions on optimized results are investigated in depth, which includes the optimized results in two-scales and the accuracy of the estimation for structural responses. Last but importantly, by comparison with the single scale optimization, the essentiality of multiscale optimization is emphasized.

In the following, the two optimization schemes are introduced in Section 2.1 and Section 2.2, respectively. For each method, the problem statement and optimization definition are first illustrated. Then, structural analysis theory as well the sensitivity analysis are given in detail. After that, several numerical examples are introduced to demonstrate the proposed schemes. This chapter is finally closed with concluding remarks in Section 2.3.

The work of this chapter is mainly related with references [26, 105].

2.1 *Concurrent multiscale stiffness optimization based on PAMP method*

2.1.1 *Problem statement of optimization problem*

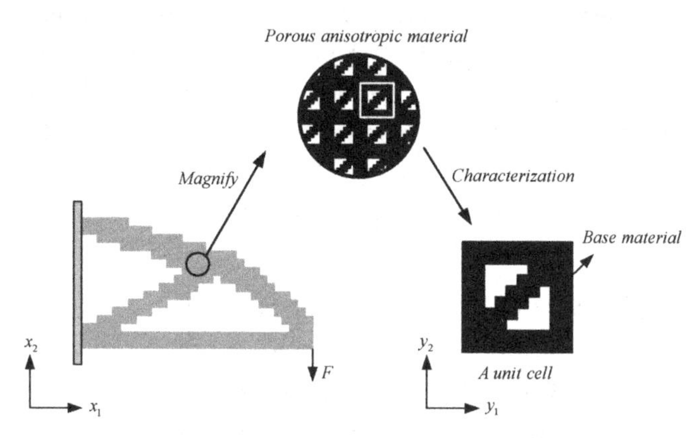

Fig. 2.1. A structure composed of porous anisotropic material with uniform microstructures

Fig. 2.1 illustrates a structure composed of a porous anisotropic material with uniform microstructures. Note that we have two materials in the following discussions: one is porous anisotropic material, and another one is base material. The base material is solid material such as aluminum, metal alloy and polymer. Porous anisotropic material, sometimes merely called 'material' for short in the following discussions, is assumed to be composed of base material and to have periodic microstructure free from any restrictions (e.g., cellular materials with square and rectangular holes, ranked laminates).

In this section, we are trying to obtain neither "black-white" design nor "varying grey" design. Our goal is to find a "single grey-white" design as shown in Fig. 2.1 with maximized structural performance by emphasizing the uniformity of the "grey level", i.e., material relative density, in macro-scale. Here arise two basic problems:

(1) Micro-scale: How to interpret the "grey" material? As stated before, the porous material can be characterized by certain periodic microstructure. So, one of the major objectives is to design the material distribution over its smallest representative unit, i.e., a unit cell (or base cell).

(2) Macro-scale: How to arrange the "grey" material? The problem in this scale lies in the optimum distribution of the "grey" material.

Both problems can be dealt with as classical layout design problems, for which topology optimization is a powerful tool. Two classes of design variables are independently defined, i.e., macro density $P(X)$ in structural design domain and micro density $\rho(Y)$ in a unit cell. Assuming optimum structural design is imbedded in the design domain Ω and is subject to external load $\boldsymbol{F}$, the minimum compliance design is formulated as

$$\text{Minimize: } C = \int_{\Omega} \boldsymbol{F}^{\mathrm{T}} \cdot \boldsymbol{U} \, d\Omega \tag{2.1}$$

$$\text{Constraint I: } \varsigma = \frac{\rho^{\mathrm{PAM}} \cdot \int_{\Omega} P \, d\Omega}{V^{\mathrm{MA}}} \leq \bar{\varsigma} \tag{2.2}$$

$$\text{Constraint II: } \rho^{\mathrm{PAM}} = \frac{\int_{Y} \rho \, dY}{V^{\mathrm{MI}}} = \overline{\varsigma^{\mathrm{MI}}} \tag{2.3}$$

$$\text{Constraint III: } 0 < \delta \leq P \leq 1, 0 < \delta \leq \rho \leq 1 \tag{2.4}$$

where C denotes structural compliance, and $\boldsymbol{U}$ represents structural displacements depending on the densities in both scales solved from the governing equation at macro scale.

Constraint I sets an upper bound on the total available base material by defining relative volume ς smaller than $\bar{\varsigma}$. V^{MA} is the area of macro design domain Ω. The density ρ^{PAM} is the specific mass density of the porous material, which is different from the relative density variable in general formulation of single-scaled topology optimization.

Constraint II makes ρ^{PAM} equal to a given value $\overline{\varsigma^{\mathrm{MI}}}$, which should be between 0.2 and 0.6 according to practical fabrication techniques. By defining this equality constraint, the material is ensured to be porous with a given relative density.

By virtue of constraints (2.2) and (2.3), the consumption of porous anisotropic material in macro-scale can thereby be predicted as

$$\frac{\varsigma}{\rho^{\mathrm{PAM}}} = \frac{\int_{\Omega} P \, d\Omega}{V^{\mathrm{MA}}} \leq \frac{\bar{\varsigma}}{\overline{\varsigma^{\mathrm{MI}}}} \tag{2.5}$$

where the left-side item of this inequality represents the consumption of porous anisotropic material in macro-scale, on which the right-side value imposes an upper bound. Since $\bar{\varsigma}$ is a given constant, bigger $\overline{\varsigma^{\mathrm{MI}}}$, i.e., more base material assigned in micro-scale design, will definitely lead to a stricter constraint for macro-scale design.

Constraint III sets bounds for density variables to avoid the singularity of stiffness matrixes in two scales, where δ is a small predetermined value that is rather close to zero.

2.1.2 *Penalization and numerical treatment related to the AH method*

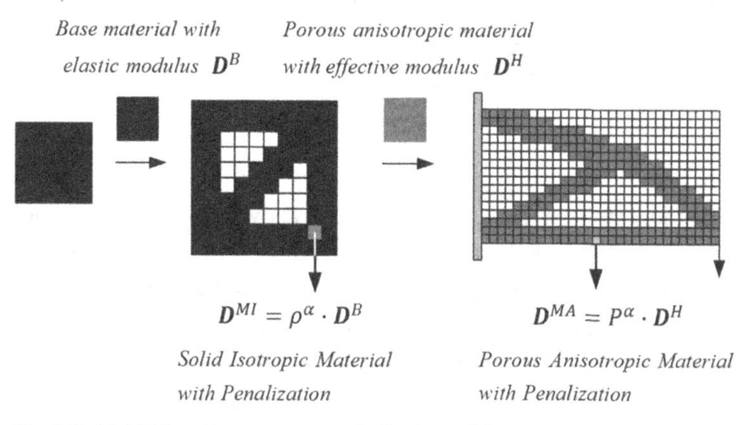

Fig. 2.2. PAMP-based concurrent optimization with two class design variables

To numerically solve the mathematical problem discretized by Finite Element Method (FEM), computational domains in both scales should be first meshed into a number of elements. As shown in Fig. 2.2, domain Ω is meshed into N elements, and domain Y is meshed into n elements. Each element is then assigned a unique density value varying between 0 and 1, i.e., P_i for the ith ($i = 1, 2, \ldots, \mathrm{N}$) element in the macroscale and ρ_j ($j = 1, 2, \ldots, \mathrm{n}$) for the jth element in the microscale, where N and n are the total number of elements in the macro and micro scales, respectively.

In order to achieve clear topologies in both scales, penalization methods are adopted. In the microscale, it is natural to utilize SIMP, a method commonly used structural topology optimization. Assuming the modulus matrix of the base material is $\boldsymbol{D}^{\mathrm{B}}$, the modulus matrix $\boldsymbol{D}^{\mathrm{MI}}$ at a micro point with density value ρ can then be expressed as

$$\boldsymbol{D}^{\mathrm{MI}} = \rho^{\alpha} \cdot \boldsymbol{D}^{\mathrm{B}} \tag{2.6}$$

where α denotes the exponent of penalization.

In the macro-scale, however, it seems no longer appropriate to use SIMP because the material here is not guaranteed to be solid isotropic. In fact, since the design of microstructure follows topology optimization procedure and no restriction is ever imposed, the chance of getting porous anisotropic material is considerably high. As a result, the macro-scale penalization named as Porous Anisotropic Material with Penalty (PAMP) is then developed, although the implementing process is remarkably similar to that of SIMP. Given porous anisotropic material with effective modulus matrix $\boldsymbol{D}^{\mathrm{H}}$, a macro point with density P has the modulus matrix $\boldsymbol{D}^{\mathrm{MA}}$ as expressed by

$$\boldsymbol{D}^{\mathrm{MA}} = P^{\alpha} \cdot \boldsymbol{D}^{\mathrm{H}} \tag{2.7}$$

Note that by introducing the relative density of the porous material as an independent macro design variable in PAMP, the macro structural topology design and material microstructure toplogy design is well decoupled. This is very different from the base-material density based multiscale optimization formulation in [25, 26], in which the macro and micro scale topology optimization are closely coupled. It makes PAMP a powerful framework for multiscale design of structure composed of structured material.

As will be indicated by numerical examples, only penalization is not enough for the microscale design. Resulting microstructural topology can be very fuzzy if no further treatment is adopted. To limit the complexity of the admissible designs and to suppress the checkerboard pattern, a number of methods have been proposed, such as enforcing an upper bound on the perimeter of the structure [106], introducing a filtering function [107] and imposing constraints on the slope of the parameters defining the geometry [108]. In this section, a variant perimeter constraint [109] will be utilized as the fourth constraint for the optimization problem.

$$\text{Constraint IV}: \gamma = \sum_{k=1}^{\mathrm{m}} l_k \cdot (\rho_{k1} - \rho_{k2})^2 \leq \bar{\gamma} \tag{2.8}$$

where m denotes the number of elements and l_k denotes the length of the k_{th} interface between elements k_1 and k_2. $\bar{\gamma}$ is a predetermined upper bound.

2.1.3 *Structural analysis and sensitivity analysis*

It is a key step in numerical optimization to establish the relationship between the objective function and design variables. Following Eq. (2.1), objective function, i.e., compliance C in the present problem, cannot be evaluated before the structural deformation $\boldsymbol{U}$ is solved. Finite element analysis is therefore formulated in macro-scale to obtain $\boldsymbol{U}$ as follows

$$\boldsymbol{K} \cdot \boldsymbol{U} = \boldsymbol{F} \tag{2.9}$$

$$\boldsymbol{K} = \int_{\Omega} \boldsymbol{B}^{\mathrm{T}} \cdot \boldsymbol{D}^{\mathrm{MA}} \cdot \boldsymbol{B} d\Omega \tag{2.10}$$

where $\boldsymbol{K}$ is the stiffness matrix of the structure, and $\boldsymbol{B}$ is the strain-displacement matrix for the macro elements. $\boldsymbol{U}$ and $\boldsymbol{F}$ are the nodal displacement and external force vector, respectively. $\boldsymbol{D}^{\mathrm{MA}}$ is defined in Eq. (2.7) as a function of $\boldsymbol{D}^{\mathrm{H}}$, which is to be determined by the following analysis.

$\boldsymbol{D}^{\mathrm{H}}$ serves as a critical link between the two-scales. On the one hand, it's a representation of effective material properties depending on microstructural configuration. On the other hand, it's also involved in macro-structural analysis as defined above. The computation of $\boldsymbol{D}^{\mathrm{H}}$ could follow the classical homogenization procedures by implementing the following two steps. Firstly, analyze the unit cell subjected to periodic boundary conditions and external forces corresponding to uniform strain fields.

$$\boldsymbol{k} \cdot \boldsymbol{u} = \int_{Y} \boldsymbol{b}^{\mathrm{T}} \cdot \boldsymbol{D}^{\mathrm{MI}} dY \tag{2.11}$$

$$\boldsymbol{k} = \int_{Y} \boldsymbol{b}^{\mathrm{T}} \cdot \boldsymbol{D}^{\mathrm{MI}} \cdot \boldsymbol{b} dY \tag{2.12}$$

where $\boldsymbol{k}$ is stiffness matrix of the microstructure, $\boldsymbol{u}$ is the microstructural displacement, $\boldsymbol{b}$ is the strain-displacement matrix and $\boldsymbol{D}^{\mathrm{MI}}$ is defined in Eq. (2.6).

Secondly, compute the effective modulus matrix by performing integration over the domain of a unit cell, i.e.,

$$\boldsymbol{D}^{\mathrm{H}} = \frac{1}{|Y|} \int_{Y} \boldsymbol{D}^{\mathrm{MI}} \cdot (\boldsymbol{I} - \boldsymbol{b} \cdot \boldsymbol{u}) dY \tag{2.13}$$

where $\boldsymbol{I}$ (3×3) is a unit matrix in two-dimensional case and $|Y|$ is the area of a unit cell.

By virtue of Eqs. (2.6)-(2.13), we have completed the structural analysis in two-scales and got the structural displacement $\boldsymbol{U}$. Based on that, one is able to obtain the current objective function C if the current design variables are available.

A typical procedure of numerical optimization generally contains two major parts: the first one is analysis, and another one is optimum search. For the latter one, a number of numerical methods are already available. In applying some derivative-based mathematical programming algorithms such as Sequential Linear Programming (SLP) and Sequential Quadratic Programming (SQP) [110], explicit expression of sensitivity is important to enhance the efficiency of the algorithm. By using all the above equations, it is readily to obtain the following two derivatives (See Appendix A for details)

$$\frac{\partial C}{\partial P_i} = -\alpha P_i^{\alpha-1} \boldsymbol{U}_i^{\mathrm{T}} \left(\int_{\Omega_i} \boldsymbol{B}^{\mathrm{T}} \cdot \boldsymbol{D}^{\mathrm{MA}} \cdot \boldsymbol{B} d\Omega_i \right) \boldsymbol{U}_i = -\frac{\alpha}{P_i} C_i \tag{2.14}$$

$$\frac{\partial C}{\partial \rho_j} = -\sum_{r=1}^{N} P_r^{\alpha} \cdot \boldsymbol{U}_r^{\mathrm{T}} \cdot \left(\int_{\Omega_r} \boldsymbol{B}^{T} \cdot \frac{\partial \boldsymbol{D}^{\mathrm{H}}}{\partial \rho_j} \cdot \boldsymbol{B} d\Omega_r \right) \cdot \boldsymbol{U}_r \tag{2.15}$$

where Ω_i is the design domain of the ith macro element, C_i is the component of the objective C associated with ith macro element, and the subscript r and i denote the quantities related to the rth and ith macro element. The derivative of $\boldsymbol{D}^{\mathrm{H}}$ with respect to ρ_j can be computed following a mapping method [111] with the results as

$$\frac{\partial \boldsymbol{D}^{\mathrm{H}}}{\partial \rho_j} = \int_{Y} (\boldsymbol{I} - \boldsymbol{b} \cdot \boldsymbol{u})^{\mathrm{T}} \cdot \frac{\partial \boldsymbol{D}^{\mathrm{MI}}}{\partial \rho_j} \cdot (\boldsymbol{I} - \boldsymbol{b} \cdot \boldsymbol{u}) dY \tag{2.16}$$

By a combination with the Eq. (2.6), the above equation leads to

$$\frac{\partial \boldsymbol{D}^{\mathrm{H}}}{\partial \rho_j} = \alpha \rho_j^{\alpha-1} \int_{Y_j} \left(\boldsymbol{I} - \boldsymbol{b} \cdot \boldsymbol{u}_j\right)^{\mathrm{T}} \cdot \boldsymbol{D}^{\mathrm{B}} \cdot \left(\boldsymbol{I} - \boldsymbol{b} \cdot \boldsymbol{u}_j\right) dY_j \tag{2.17}$$

Now, we have shown structural analysis and sensitivity analysis. Fig. 2.3 gives a detailed flow chart for structural analysis and optimization based on AH method with each key step associated with corresponding equations above.

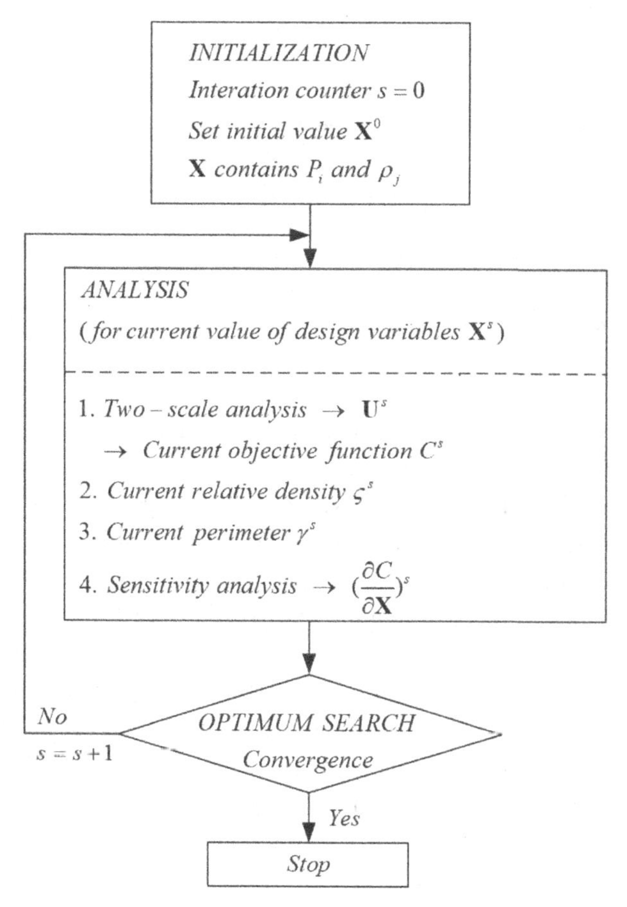

Fig. 2.3. Flow chart of the two-scale concurrent optimization based on the AH method

2.1.4 *Numerical examples*

(1) **Example 1:** A discussion on effect of parameters $\bar{\varsigma}$, $\overline{\varsigma^{MI}}$ and $\bar{\gamma}$

We first consider a MBB beam. The main objective is both to illustrate the proposed method and to discuss the parameters $\bar{\varsigma}$, $\overline{\varsigma^{\mathrm{MI}}}$, and $\bar{\gamma}$. As will be demonstrated, all of the three parameters do influence the resulting topologies (structure and microstructure) and compliance. But they are different in nature: $\bar{\varsigma}$ and $\overline{\varsigma^{MI}}$ are involved in physical problem as limits of material consumption, while $\bar{\gamma}$ is merely a controlling parameter in numerical treatment.

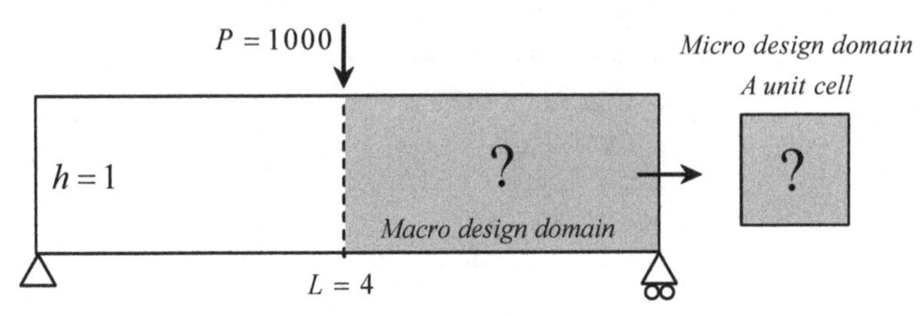

Fig. 2.4. Illustration of MBB beam used for the optimization of Example 1 based on the AH method

A MBB beam in Fig. 2.4 is loaded with a concentrated vertical force of $P = 1000$ at the center of the top edge and is supported on rollers at the bottom-right corner and on fixed supports at the bottom-left corner. The base material is assumed to have Young's modulus $E = 2.1 \times 10^5$ and Poisson's ratio $\upsilon = 0.3$. Geometric parameters are $L = 4$ and $h = 1$. As we are only interested in qualitative results, the dimensions and load for this problem are chosen non-dimensional. Due to the symmetry of the problem, only the right half part is considered as macro design domain. The mesh is 50×25 for macro design domain and 25×25 for the microstructure (8-node bilinear plane element). Optimum search is implemented by the SQP algorithm.

To discuss the influence of the adopted perimeter constraint, the optimization problem is first solved with fixed $\bar{\varsigma}$=0.1, $\overline{\varsigma^{MI}}$=0.4 and varying perimeter constraint $\bar{\gamma}$. As shown in Table 2.1, if perimeter constraint is not applied or the controlling parameter is not low enough, the resulting microstructural topology turns out to be very complex, and the compliance value is relatively higher than those with appropriate $\bar{\gamma}$. Consequently, it is suggested, partly based on the numerical results, that it is necessary to apply numerical treatments in microscale design to ensure a clear topology and better system performance. However, note that too small $\bar{\gamma}$ may lead to difficulties of convergence.

Table 2.1. Optimized results for varying $\bar{\gamma}$

$\bar{\varsigma}$	$\overline{\varsigma^{MI}}$	$\bar{\gamma}$	Compliance	Microstructural topology
0.12	0.4	2	7515	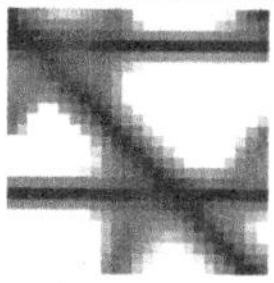
0.12	0.4	3	5675	
0.12	0.4	4	5676	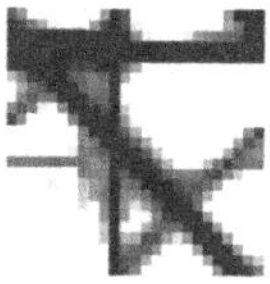
0.12	0.4	8	7077	
0.12	0.4	N/A	7583	

Then, we will set $\bar{\gamma}$ to an appropriate value of 4 and solve the problem again with varying $\bar{\varsigma}$. As shown in Table 2.2, although resulting microstructures are all porous anisotropic, the topology does vary considerably for different cases. When $\bar{\varsigma} = 0.25$, the microstructure after a rotation of 45 degrees closely resembles the so-called 'Triangular cell' (see Fig. 2.5), which is considered to have superior in-plane mechanical properties [112]. And when $\bar{\varsigma} = 0.075$, the microstructure after a rotation of 45 degrees is somewhat more similar to the 'Mixed cell' (see Fig. 2.5), which is considered as another competitive microstructure for linear cellular material. The resemblance between optimized results and some existing superior microstructures implies that the proposed method does generate optimum micro-topology. In the macro-scale, the design also changes with the increase of $\bar{\varsigma}$ in accordance with different material properties and changing available porous materials. As for the compliance value, it is reasonable that better system performance is achieved with a more available base material.

Table 2.2. Optimized results for varying $\bar{\varsigma}$

$\bar{\varsigma}$	$\overline{\varsigma^{MI}}$	Compliance	Structural topology	Microstructural topology
0.075	0.4	9855	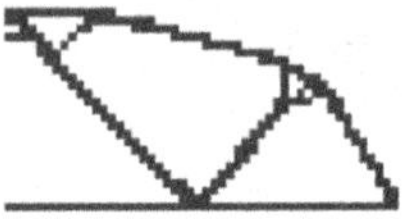	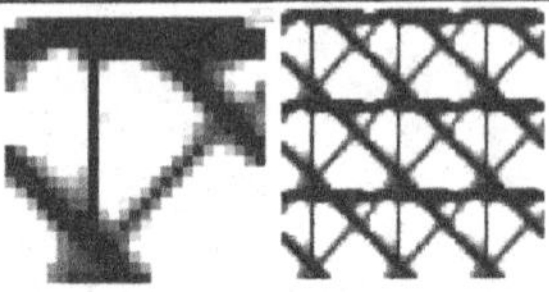
0.09	0.4	7292	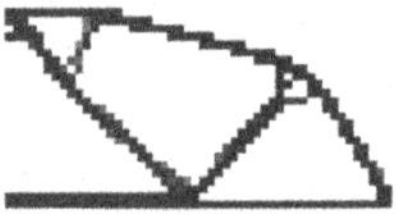	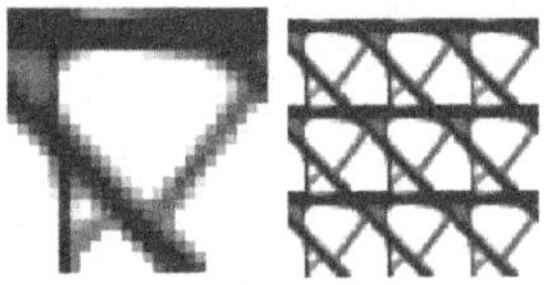
0.12	0.4	5676	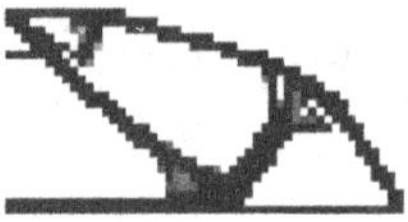	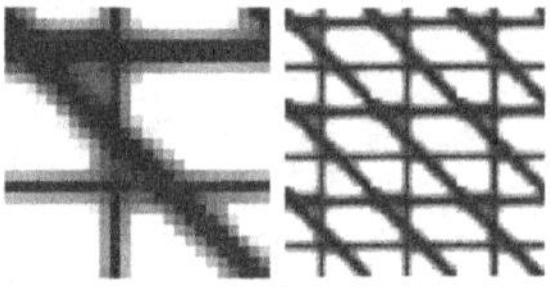
0.18	0.4	3707		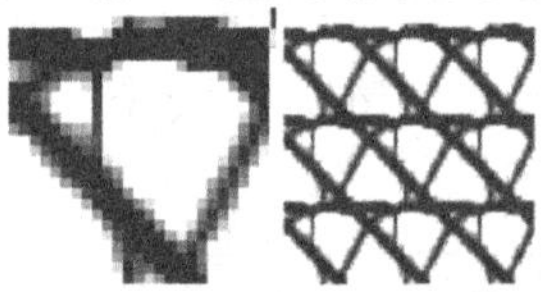
0.25	0.4	2234		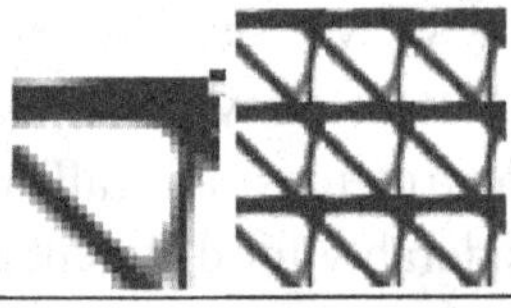

As aforementioned, there exists a strong trade-off between macro-scale design and micro-scale design, namely the allocation of material between the two-scales which can be characterized by $\bar{\varsigma}/\overline{\varsigma^{MI}}$ and $\overline{\varsigma^{MI}}$. To illustrate the discussion, the two-scale design problem will be solved again with predetermined available base material $\bar{\varsigma} = 0.12$ and varying $\overline{\varsigma^{MI}}$. As shown in Table 2.3, a larger value of $\overline{\varsigma^{MI}}$ leads to a lower system compliance with more base material consumption in micro-scale design and less porous material consumption in the macro-scale design. This tendency suggests that the structure made of solid material is stiffer than the one made of porous material if the total material is the same and both macro-structural and microstructural topology are free to be optimized. The usage of porous material in multiphysics environment should be preferred. The more discussions can be found in Chapter 5.

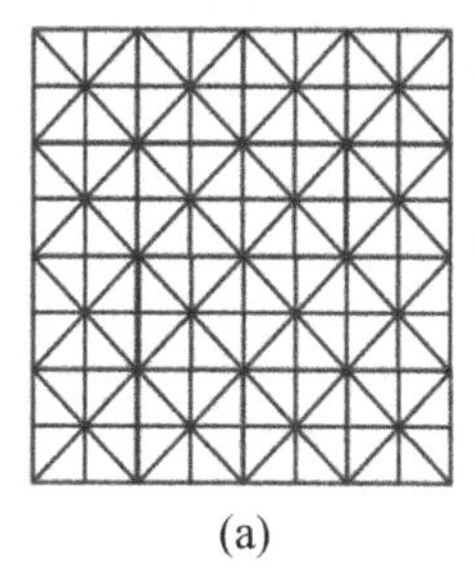

(a)

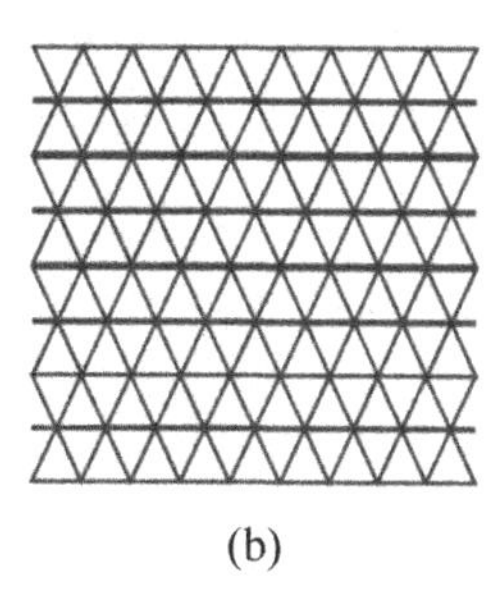

(b)

Fig. 2.5. The microstructures for linear cellular material [112]. (a) Mixed cell. (b) Triangular cell

Table 2.3. Optimized results for varying $\overline{\varsigma^{MI}}$

$\bar{\varsigma}$	$\overline{\varsigma^{MI}}$	Compliance	Structural topology	Microstructural topology
0.12	0.2	8880		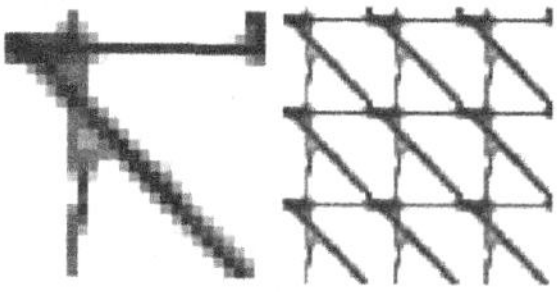
0.12	0.3	6210	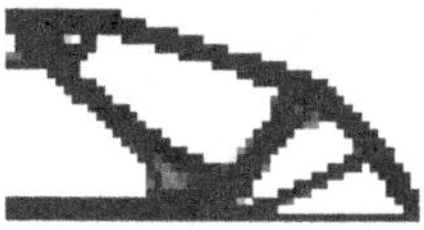	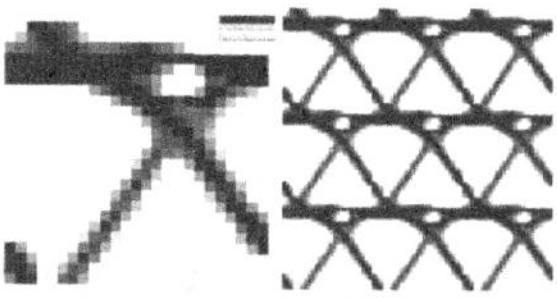
0.12	0.4	5676	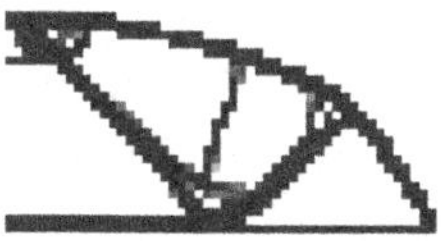	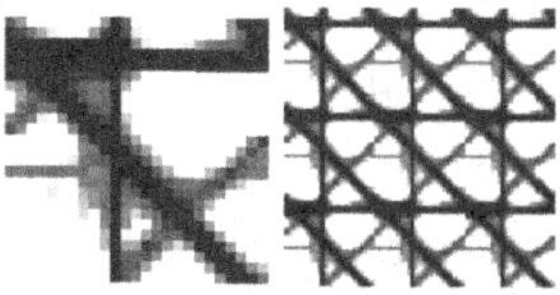
0.12	0.5	5487	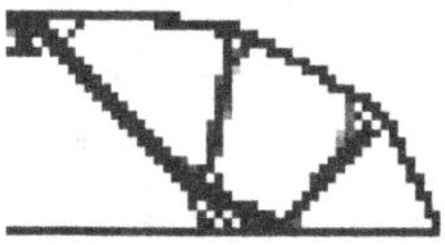	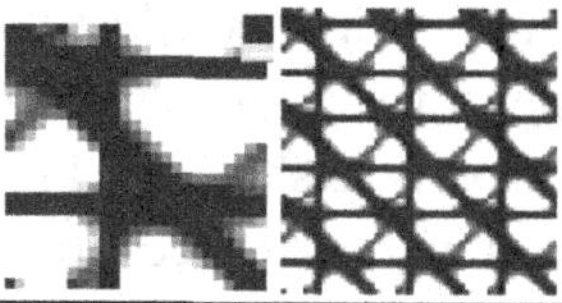

(2) **Example 2:** A validation of the proposed method

An L-shaped beam in Fig. 2.6 is considered in this example to validate the proposed method by comparing numerical results with expectations. The upper edge is fixed, and a linearly varying distributed loads is applied along the right edge to simulate a moment. As for the geometry, $a = 1$ and b is set as a variable determined by λ. Constraint parameters $\bar{\varsigma}$, $\overline{\varsigma^{\mathrm{MI}}}$ and $\bar{\gamma}$ are respectively fixed to be 0.1, 0.4, and 4.

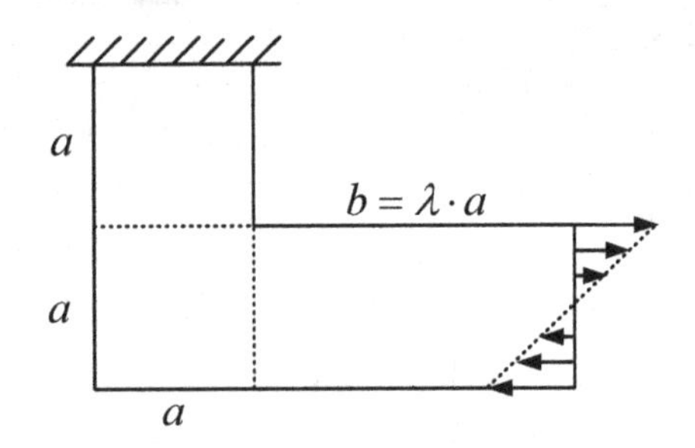

Fig. 2.6. An L-shaped beam

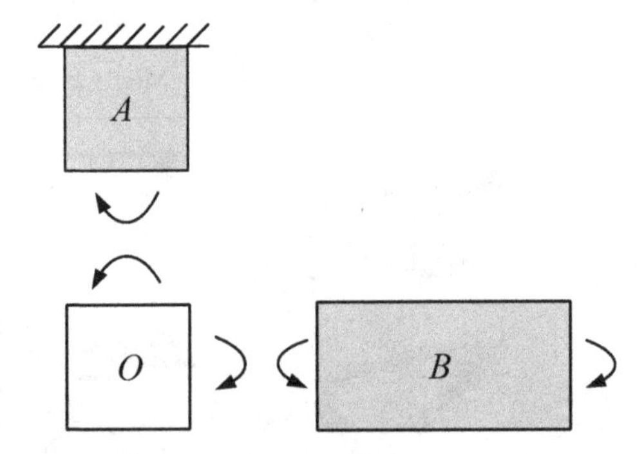

Fig. 2.7. A decomposition of the L-shaped beam

Note that within the whole structure, the material is assumed uniform, but stress distributions could vary considerably, so the design in the microscale is actually a multi-objective optimization problem. This example is a good illustration for the concurrent multiscale optimization design of structures/materials. The L-shaped beam can be divided into three parts as shown in Fig. 2.7. For part B, the moments are imposed on its left and right faces resulting in principal stress in x-direction, and we can, therefore, expect a larger effective Young's modulus in x-direction for material design. Following similar analysis, the microstructural topology should be stiffer in y-direction for part A. So, the material designs of the two parts seem to be contradictory to some extent. However, they must be included in one optimization problem due to the assumption of material uniformity.

This contradiction makes it important the weight between the two parts, and λ is defined to adjust the weight. Four cases with different λ (0, 1, 2, and 3) are respectively considered. If $\lambda = 0$, part B does not exist, and part A is dominant; If $\lambda = 1$, the weight of part B is increased; If $\lambda = 2$ and 3, part B gradually becomes the dominance.

Table 2.4. Optimized results of the L-shaped beam design

$\bar{\varsigma}$	$\overline{\varsigma^{\mathrm{MI}}}$	λ	Compliance	D_{11}^H / D_{22}^H	Structural topology	Microstructural topology
0.1	0.4	0	43012	0.55		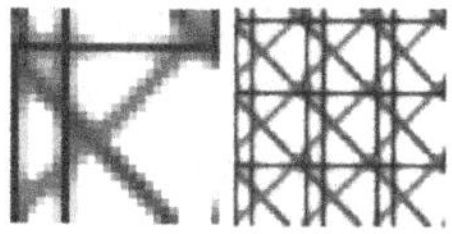
0.1	0.4	1	56958	1.38		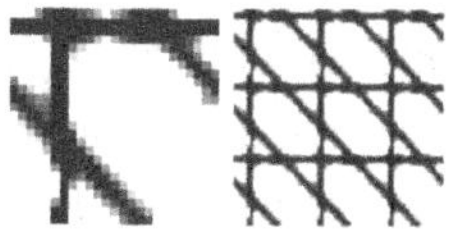
0.1	0.4	2	64345	1.08		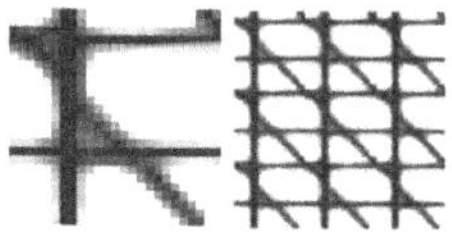
0.1	0.4	3	76590	1.27	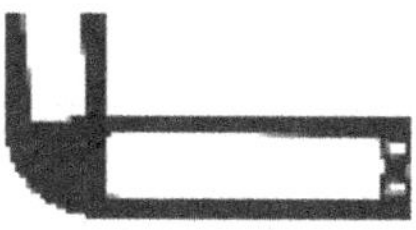	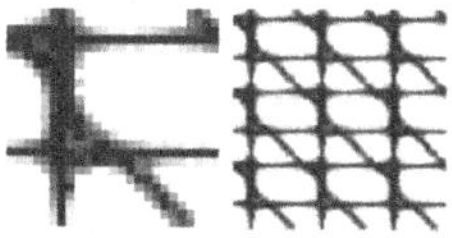

The optimized results considering different λ are listed in Table 2.4. Modulus ratio $D_{11}^{\mathrm{H}}/D_{22}^{\mathrm{H}}$ is introduced to indicate the contrast of material arrangement in the two directions where D_{pq}^{H} (p, q=1, 2, 3 in 2D problem) denotes the modulus matrix of an element. As shown in Table 2.4, the modulus ratio is increased with increasing λ, which means base material is increasingly arranged to strengthen the porous material in x-direction. This result is perfectly consistent with the above theoretical analysis, which again verifies the micro-scale design in a qualitative sense.

2.2 *Concurrent multiscale stiffness optimization based on Extended Multiscale Finite Element Method (EMsFEM)*

Lattice material is a special type of porous material and its microstructure is usually modeled as a truss unit cell in the structural analysis and optimization. Since the practical structure may include hundred million truss bars, direct numerical simulation (DNS) of structure made of lattice material is highly time consuming. Therefore, structural and material optimization based on DNS is impossible. In Section 2.1, we apply AH method to relate the micro structure of the unit cell and the macro structure. However, the method suffers from size effect. In this section, we will show an alternative approach, that is, the concurrent multiscale analysis and optimization approach based on EMsFEM.

As shown in Fig. 2.8, the macrostructure is discreated into N elements, and each element are constructed by a uniform unit cell, where M trusses are included. The concurrent multiscale stiffness optimization problem based on PAMP is formulated as follows

$$\text{Minimize: } C = \int_{\Omega} \boldsymbol{F}^{\mathrm{T}} \cdot \boldsymbol{U}\, d\Omega \tag{2.18}$$

$$\text{Constraint I: } \varsigma = \frac{\rho^{\mathrm{PAM}} \cdot \int_{\Omega} \rho\, d\Omega}{V^{\mathrm{MA}}} \leq \bar{\varsigma} \tag{2.19}$$

$$\text{Constraint II: } \rho^{\mathrm{PAM}} = \frac{\sum_{i=1}^{\mathrm{N}} A_i L_i}{V^{\mathrm{MI}}} = \overline{\varsigma^{\mathrm{MI}}} \tag{2.20}$$

$$\text{Constraint III: } A_{\mathrm{L}} \leq A_i \leq A_{\mathrm{U}}, 0 < \delta \leq \rho_j \leq 1 \tag{2.21}$$

where L_i and A_i are length and cross-sectional area of the ith bar, respectively, where $i = 1,2,\ldots,\mathrm{M}$. A_{L} and A_{U} represent the lower and upper bound of the cross-sectional area, respectively. $A_{\mathrm{L}} = 0.001$ and $A_{\mathrm{U}} = 100$ will be employed in the following numerical examples. ρ_j represents the density of the jth macro element and $j = 1,2,\ldots,\mathrm{N}$. $\boldsymbol{U}$ represents structural displacements depending on the design variables in both scales solved from the EMsFEM. The rest quantities are identical to that in Section 2.1.

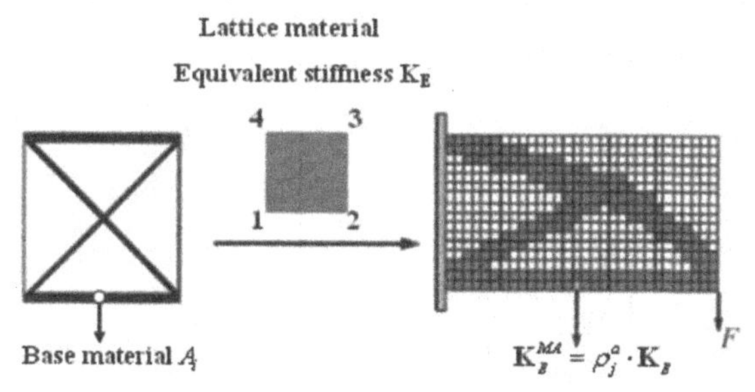

Fig. 2.8. Concurrent multiscale optimization of lattice materials based on EMsFEM

2.2.1 *Basic concepts of Extended Multiscale Finite Element Method*

In the traditional FEM method, the material in each element is assumed to be homogeneous and the element stiffness matrix is constructed based on assumed analytic shape function. The EMsFEM proposed in Zhang et al. [6] extends the FEM method to multiscale analysis of the structure made of heterogeneous material. In the EMsFEM, the structure is discretized by a number of macro elements and material in macro element can be heterogeneous. And the stiffness matrix of each macro element is constructed in a similar way with the traditional FEM but based on numerical shape function, which is obtained by applying microscale FEM analysis to each macro element. To enhance the accuracy of the element stiffness matrix for heterogeneous material element, the additional coupling items in the shape function are introduced in EMsFEM. Here, the macro elements modeled by a truss unit cell is considered. As shown in Fig. 2.9, taking a four-node truss unit cell as example, the nodal displacement of the microstructure in the macro element can be expressed as

$$\boldsymbol{u} = \boldsymbol{N}\boldsymbol{u}'_{\mathrm{E}} \tag{2.22}$$

where $\boldsymbol{N}$ is the multiscale shape function matrix of the truss unit cell, $\boldsymbol{u}$ and $\boldsymbol{u}'_{\mathrm{E}}$ denote the displacement vectors of all the nodes in the microscale and the four corner nodes in the macro-scale, respectively. And they can be written as

$$\boldsymbol{u} = [u_1\ v_1\ u_2\ v_2\ \ldots\ldots\ u_n\ v_n]^{\mathrm{T}} \tag{2.23}$$

$$\boldsymbol{N} = \left[\boldsymbol{R}_{\mathrm{x}}(1)^{\mathrm{T}}\ \boldsymbol{R}_{\mathrm{y}}(1)^{\mathrm{T}}\ \boldsymbol{R}_{\mathrm{x}}(2)^{\mathrm{T}}\ \boldsymbol{R}_{\mathrm{y}}(2)^{\mathrm{T}}\ \ldots\ldots\ \boldsymbol{R}_{\mathrm{x}}(n)^{\mathrm{T}}\ \boldsymbol{R}_{\mathrm{y}}(n)^{\mathrm{T}}\right] \tag{2.24}$$

$$\boldsymbol{u}'_{\mathrm{E}} = [u'_1\ v'_1\ u'_2\ v'_2\ u'_3\ v'_3\ u'_4\ v'_4]^{\mathrm{T}} \tag{2.25}$$

$$\boldsymbol{R}_{\mathrm{x}}(i) = \begin{bmatrix} N_{1xx}(i) \\ N_{1xy}(i) \\ N_{2xx}(i) \\ N_{2xy}(i) \\ N_{3xx}(i) \\ N_{3xy}(i) \\ N_{4xx}(i) \\ N_{4xy}(i) \end{bmatrix}^{\mathrm{T}} \quad \boldsymbol{R}_{\mathrm{y}}(i) = \begin{bmatrix} N_{1yx}(i) \\ N_{1yy}(i) \\ N_{2yx}(i) \\ N_{2yy}(i) \\ N_{3yx}(i) \\ N_{3yy}(i) \\ N_{4yx}(i) \\ N_{4yy}(i) \end{bmatrix}^{\mathrm{T}} \tag{2.26}$$

where n is the number of micro nodes in the truss unit cell, $N_{jyx}(i)$ ($j = 1,2,3,4; i = 1,2,\ldots,n$) is the coupling additional item of the shape function. Its physical meaning refers to the nodal displacements in y-direction of the ith micro nodes when the node j of the macro element has a unit displacement in the x-direction.

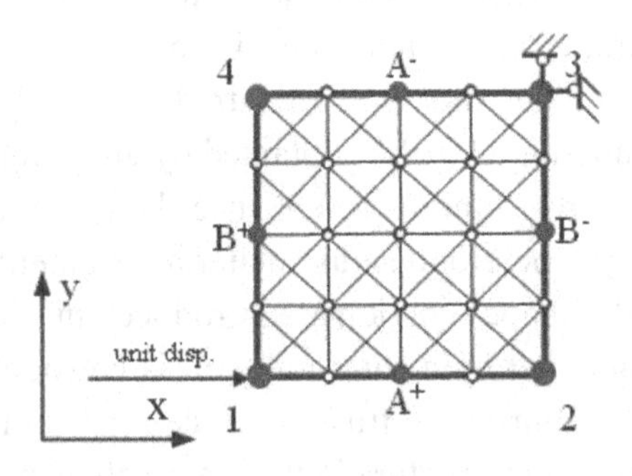

Fig. 2.9. A unit cell modeled by truss elements

The macro element stiffness matrix $\boldsymbol{K}_{\mathrm{E}}$ of the truss unit cell with EMsFEM according to strain energy equivalence can be described as

$$\boldsymbol{K}_{\mathrm{E}} = \sum_{i=1}^{\mathrm{M}} (\boldsymbol{R}^{\mathrm{e}i})^{\mathrm{T}} (\boldsymbol{\theta}^{\mathrm{e}i})^{\mathrm{T}} k_{\mathrm{e}}^{i} \boldsymbol{\theta}^{\mathrm{e}i} \boldsymbol{R}^{\mathrm{e}i} \tag{2.27}$$

where M is the total number of rods within the truss unit cell, $\boldsymbol{R}^{ei}$ is the numercal shape function of ith rod which can be determined by Eqs. (2.24) and (2.26), k_{e}^{i} is coefficient of elasticity of i th rod, $\boldsymbol{\theta}^{ei}$ is the trigonometric functions matrix between ith rod and x-axis.

Once the elemental stiffness matrix $\boldsymbol{K}_{\mathrm{E}}$ is obtained, macro structure, even with complex geometry and loading/boundary conditions, can be discretized by the four-node planar finite element with EMsFEM. Then we can use the classical finite element technique to obtain the nodal displacements $\boldsymbol{u}_{\mathrm{E}}'$ of four-node element in the macroscale. By substituting the displacement results into Eq. (2.22) to implement downscaling calculation, the nodal displacement $\boldsymbol{u}$ of truss unit cell in the microscale is thereby obtained. Further, the stress in micro bar can be calculated.

2.2.2 *Discussions of boundary conditions*

From the above derivations, it can be found that one critical step of EMsFEM is to obtain the numerical shape function for the unit cell. This can be done by solving the displacement field for the unit cell under proper boundary conditions. Since the construction of the shape functions influents the accuracy of analysis with EMsFEM directly, two boundary conditions, i.e., the linear and periodic boundary conditions are tested blew.

Reusing the unit cell shown in Fig. 2.9 as an example, for linear boundary conditions, the shape function $\boldsymbol{N}_x^1 = \left[N_{1xx}\ N_{1yx}\right]$ can be obtained by applying a unit displacement in x-direction to node 1 according to the physical meaning of the shape functions. The boundary 43 and 23 are fixed to avoid the rigid displacement. The x displacements along the boundary 12 and 14 are linearly varying from 1 to 0, and their y displacements are constrained simultaneously. For periodic boundary conditions, node 3 is fixed. Correspondingly, micro nodes $(A^+,\ A^-)$ and $(B^+,\ B^-)$ on boundaries meet the following relations, respectively

$$\begin{cases} u^{A^+} - u^{A^-} = \Delta x \\ v^{A^+} = v^{A^-} \end{cases} \begin{cases} u^{B^+} - u^{B^-} = \Delta y \\ v^{B^+} = v^{B^-} \end{cases} \tag{2.28}$$

where Δx and Δy are linearly varying from 1 to 0 along the boundary 12 and 14. It implies that the displacements of the corner nodes 2, 3 and 4 are 0. Then, $\boldsymbol{N}_x^1$ under the two boundary conditions can be obtained by solving the micro analysis problems defined above, respectively. The rest of the shape functions can be constructed in a similar way.

2.2.3 *Structural analysis and sensitivity analysis*

The problem statement is similar as that in Section 2.1.1. Structural analysis for the lattice materials and structures is carried out with EMsFEM described in the previous section. The detailed formula can be written as follows

$$\boldsymbol{K}^{\mathrm{MA}} \cdot \boldsymbol{U} = \boldsymbol{F} \tag{2.29}$$

$$\boldsymbol{K}^{\mathrm{MA}} = \sum_{j=1}^{\mathrm{N}} \boldsymbol{T}^{\mathrm{T}} \cdot \rho_j^{\alpha} \left(\sum_{i=1}^{\mathrm{M}} \left(\boldsymbol{R}^{\mathrm{e}i}\right)^{\mathrm{T}} \left(\boldsymbol{\theta}^{\mathrm{e}i}\right)^{\mathrm{T}} k_{\mathrm{e}}^{i} \boldsymbol{\theta}^{\mathrm{e}i} \boldsymbol{R}^{\mathrm{e}i} \right) \cdot \boldsymbol{T} \tag{2.30}$$

where $\boldsymbol{K}^{\mathrm{MA}}$ is the general stiffness matrix of the macro structure constituted by the elemental stiffness matrix $\boldsymbol{K}_{\mathrm{E}}$. N is the number of four-node elements in macro design domain, M is the number of the rods in the micro truss unit cell, ρ_j is the relative density of the jth four-node element, α is the penalty coefficient, and $\boldsymbol{T}$ is transformation matrix. Then we have the completed analysis formula of the structure composed of lattice materials on the two-scales.

The algorithm SQP is also used as the optimization solver. The sensitivity analysis is carried out to the design variables on the two geometrical scales, respectively.

Derivative of the objective function to the macro-design variable ρ_j is

$$\begin{aligned} \frac{\partial C}{\partial \rho_j} &= -\boldsymbol{U}^{\mathrm{T}} \cdot \frac{\partial \boldsymbol{K}^{\mathrm{MA}}}{\partial \rho_j} \cdot \boldsymbol{U} \\ &= -\boldsymbol{U}_j^{\mathrm{T}} \cdot \alpha \rho_j^{\alpha-1} \cdot \left(\sum_{i=1}^{\mathrm{M}} \left(\boldsymbol{R}^{\mathrm{e}i}\right)^{\mathrm{T}} \left(\boldsymbol{\theta}^{\mathrm{e}i}\right)^{\mathrm{T}} k_{\mathrm{e}}^{i} \boldsymbol{\theta}^{\mathrm{e}i} \boldsymbol{R}^{\mathrm{e}i} \right) \cdot \boldsymbol{U}_j, \end{aligned} \tag{2.31}$$

and derivative of the objective function to the micro-design variable A_i can be expressed as

$$
\frac{\partial C}{\partial A_i} = -\boldsymbol{U}^{\mathrm{T}} \cdot \frac{\partial \boldsymbol{K}^{\mathrm{MA}}}{\partial A_i} \cdot \boldsymbol{U}
$$

$$
= -\sum_{r=1}^{\mathrm{N}} \boldsymbol{U}_r^{\mathrm{T}} \cdot \rho_r^{\alpha} \cdot \left(\frac{\partial}{\partial A_i} \sum_{j=1}^{\mathrm{M}} \left(\boldsymbol{R}^{\mathrm{e}j}\right)^{\mathrm{T}} \left(\boldsymbol{\theta}^{\mathrm{e}j}\right)^{\mathrm{T}} k_{\mathrm{e}}^{j} \boldsymbol{\theta}^{\mathrm{e}j} \boldsymbol{R}^{\mathrm{e}j} \right) \cdot \boldsymbol{U}_r
$$

$$
= -\sum_{r=1}^{\mathrm{N}} \boldsymbol{U}_r^{\mathrm{T}} \cdot \rho_r^{\alpha} \cdot \left(\begin{array}{c} \sum_{j=1}^{\mathrm{M}} \frac{\partial \left(\boldsymbol{R}^{\mathrm{e}j}\right)^{\mathrm{T}}}{\partial A_i} \left(\boldsymbol{\theta}^{\mathrm{e}j}\right)^{\mathrm{T}} k_{\mathrm{e}}^{j} \boldsymbol{\theta}^{\mathrm{e}j} \boldsymbol{R}^{\mathrm{e}j} \\ + \left(\boldsymbol{R}^{\mathrm{e}j}\right)^{\mathrm{T}} \left(\boldsymbol{\theta}^{\mathrm{e}j}\right)^{\mathrm{T}} \cdot \frac{E_i}{L_i} \cdot \boldsymbol{\theta}^{\mathrm{e}j} \boldsymbol{R}^{\mathrm{e}j} \\ + \sum_{j=1}^{\mathrm{M}} \left(\boldsymbol{R}^{\mathrm{e}j}\right)^{\mathrm{T}} \left(\boldsymbol{\theta}^{\mathrm{e}j}\right)^{\mathrm{T}} \cdot k_{\mathrm{e}}^{j} \cdot \boldsymbol{\theta}^{\mathrm{e}j} \frac{\partial \boldsymbol{R}^{\mathrm{e}j}}{\partial A_i} \end{array} \right) \cdot \boldsymbol{U}_r \tag{2.32}
$$

where the subscript j and r denote the quantities related to the jth and rth macro element. From the above expressions, the derivative of the objective function to the macro design variables ρ_i is relatively easy to obtain. However, the key of sensitivity analysis in micro-scale is to solve the derivative of $\boldsymbol{R}^e$ to the micro design variables A_i. According to Eqs. (2.22) and (2.24), the problem can be transformed to seek the derivatives of the nodal displacement of the truss unit cell to the micro design variables A_i under a specified displacement boundary condition.

For the linear boundary condition, the finite element formulation in micro-scale can be expressed as

$$
\overline{\boldsymbol{K}}^{\mathrm{MI}} \cdot \overline{\boldsymbol{u}} = \overline{\boldsymbol{p}} \tag{2.33}
$$

Solving the derivative of both sides of Eq. (2.33)

$$
\frac{\partial \overline{\boldsymbol{u}}}{\partial A_i} = \left(\overline{\boldsymbol{K}}^{\mathrm{MI}}\right)^{-1} \cdot \frac{\partial \overline{\boldsymbol{p}}}{\partial A_i} - \left(\overline{\boldsymbol{K}}^{\mathrm{MI}}\right)^{-1} \cdot \frac{\partial \overline{\boldsymbol{K}}^{\mathrm{MI}}}{\partial A_i} \cdot \overline{\boldsymbol{u}} \tag{2.34}
$$

where, $\overline{\boldsymbol{K}}^{\mathrm{MI}}$ is the stiffness matrix of truss unit cell, $\overline{\boldsymbol{u}}$ and $\overline{\boldsymbol{p}}$, respectively, are the nodal displacement vector within the truss unit cell and the force vector by applying with the linear boundary condition. For those boundary nodes with the specified displacements, the derivative of the nodal displacement to the design variable A_i is 0, because it is not altered with the change of design variable A_i. Thus, the derivatives of all the nodal displacements of the truss unit cell to the micro design variable A_i can be obtained.

For the periodic boundary condition, the finite element formula can be described using the similar manner as Eq. (2.33). Considering the periodic conditions (2.28) of the nodal displacements on the boundary, the nodes of the truss unit cell can be grouped into the master nodes, slave nodes and internal nodes. Take the truss unit cell in Fig. 2.9 as an example, the nodes on the boundaries 2-3, 4-3 are master nodes, the nodes on the 1-2 and 1-4 are slaves nodes. The displacement of the slave nodes is controlled by the corresponding master nodes through the periodic conditions. In this way the nodal displacement vector of the truss unit cell can be expressed as

$$\widehat{\boldsymbol{K}}^{\mathrm{MI}} \cdot \widehat{\boldsymbol{u}} = \widehat{\boldsymbol{p}} \tag{2.35}$$

$$\widehat{\boldsymbol{u}} = \left[\widehat{\boldsymbol{u}}_{\mathrm{m}}^{\mathrm{T}}\ \widehat{\boldsymbol{u}}_{\mathrm{s}}^{\mathrm{T}}\ \widehat{\boldsymbol{u}}_{\mathrm{i}}^{\mathrm{T}}\right]^{\mathrm{T}} \tag{2.36}$$

$$\widehat{\boldsymbol{u}}_{\mathrm{s}} = \widehat{\boldsymbol{u}}_{\mathrm{m}} + \Delta \tag{2.37}$$

where $\widehat{\boldsymbol{u}}_{\mathrm{m}}$ is the displacement vector of master nodes on boundaries, $\widehat{\boldsymbol{u}}_{\mathrm{s}}$ is the displacement vector of slave nodes on boundaries, $\widehat{\boldsymbol{u}}_{\mathrm{i}}$ is the displacement vector of the internal nodes, Δ is a constant vector which is changed between 0 and 1 along the boundary of the truss unit cell linearly. Substitute Eqs. (2.36) and (2.37) into Eq. (2.35) and remove the displacement vector of slave nodes from the unknown variales $\widehat{\boldsymbol{u}}$, Eq. (2.35) can be written as

$$\begin{bmatrix} \widehat{\boldsymbol{K}}_{\mathrm{mm}} + \widehat{\boldsymbol{K}}_{\mathrm{ms}} + \widehat{\boldsymbol{K}}_{\mathrm{sm}} + \widehat{\boldsymbol{K}}_{\mathrm{ss}} & \widehat{\boldsymbol{K}}_{\mathrm{mi}} + \widehat{\boldsymbol{K}}_{\mathrm{si}} \\ \widehat{\boldsymbol{K}}_{im} + \widehat{\boldsymbol{K}}_{is} & \widehat{\boldsymbol{K}}_{ii} \end{bmatrix} \cdot \begin{bmatrix} \widehat{\boldsymbol{u}}_{\mathrm{m}} \\ \widehat{\boldsymbol{u}}_{\mathrm{i}} \end{bmatrix} = \begin{bmatrix} \widehat{\boldsymbol{p}}_{\mathrm{m}} - (\widehat{\boldsymbol{K}}_{\mathrm{ms}} + \widehat{\boldsymbol{K}}_{\mathrm{ss}})\Delta \\ \widehat{\boldsymbol{p}}_{\mathrm{i}} - \widehat{\boldsymbol{K}}_{\mathrm{is}}\Delta \end{bmatrix} \tag{2.38}$$

Notably, Eq. (2.38) can be simply expressed as

$$\overline{\overline{\boldsymbol{K}}}^{\mathrm{MI}} \cdot \overline{\overline{\boldsymbol{u}}} = \overline{\overline{\boldsymbol{p}}} \tag{2.39}$$

where $\overline{\overline{\boldsymbol{K}}}^{\mathrm{MI}}$ is the stiffness matrix of truss unit cell, $\overline{\overline{\boldsymbol{u}}}$ and $\overline{\overline{\boldsymbol{p}}}$, respectively, are the nodal displacement vector within the truss unit cell and the force vector by applying with the periodic boundary. Solving the derivatives of both sides of Eq. (2.39)

$$\frac{\partial \overline{\overline{\boldsymbol{u}}}}{\partial A_i} = \left(\overline{\overline{\boldsymbol{K}}}^{\mathrm{MI}}\right)^{-1} \cdot \frac{\partial \overline{\overline{\boldsymbol{p}}}}{\partial A_i} - \left(\overline{\overline{\boldsymbol{K}}}^{\mathrm{MI}}\right)^{-1} \cdot \frac{\partial \overline{\overline{\boldsymbol{K}}}^{\mathrm{MI}}}{\partial A_i} \cdot \overline{\overline{\boldsymbol{u}}} \tag{2.40}$$

Thus, the derivatives of nodal displacements of master nodes and internal nodes to design variable A_i are obtained. The derivatives of nodal displacements of salve nodes are the same as their counterparts of master nodes. And the derivatives of nodal displacements with specified deformation are 0. Then the derivatives of all the nodal displacement within the truss unit cell are solved.

The derivatives of constraint I to the macro and micro design variables are

$$\frac{\partial \varsigma}{\partial A_i} = \frac{L_i \cdot \int_{\Omega} \rho \, d\Omega}{V^{\mathrm{MI}} V^{\mathrm{MA}} \bar{\varsigma}} \tag{2.41}$$

$$\frac{\partial \varsigma}{\partial \rho_j} = \frac{\rho^{\mathrm{PAM}} \cdot A_j^{\mathrm{MA}}}{V^{\mathrm{MA}} \bar{\varsigma}} \tag{2.42}$$

where A_j^{MA} is the area of j_{th} four-node element in the macro-design domain, L_i is the length of i-th rod within the truss unit cell. The derivatives of constraint II to the macro and micro design variables are

$$\frac{\partial \rho^{\mathrm{PAM}}}{\partial A_i} = \frac{L_i}{V^{\mathrm{MI}} \overline{\varsigma^{\mathrm{MI}}}} \tag{2.43}$$

$$\frac{\partial \rho^{PAM}}{\partial \rho_j} = 0 \tag{2.44}$$

Now we have completed the structural analysis and sensitivity analysis. The major numerical procedures can be briefly summarized as

(1) Initialize micro and macro design variables (A_i and ρ_i).

(2) Compute the penalized macro physical densities of each element in the macro-scale.

(3) Use EMsFEM for the truss unit cell structural analysis with linear or periodic boundary condition in micro-scale to obtain the elemental stiffness matrix $\boldsymbol{K}_{\mathrm{E}}$ in Eq. (2.27).

(4) Solve the macro nodal displacement $\boldsymbol{U}$ using Eq. (2.29).

(5) Obtain the value of objective function C in Eq. (2.1).

(6) Compute constraints: total volume fraction ς and relative density ρ^{PAM} of truss unit cell in Eqs. (2.2) and (2.3), respectively.

(7) Compute the sensitivities of the objective and constraints with respect to the design variables using Eqs. (2.31)-(2.32) and Eqs. (2.41)-(2.44), respectively.

(8) Adopt SQP for optimum search and update the design variables. If convergence, go to step (9). Otherwise, go to step 2.

(9) Output the final topology result and Stop.

2.2.4 *Numerical examples*

For the convenience of calculation and comparison, the parameters of optimization examples are dimensionless, and the elastic modulus of the base material is taken as 10^6. In the following examples, the linear and periodic boundary conditions are both considered.

(1) **Example 1:** A discussion of size effect

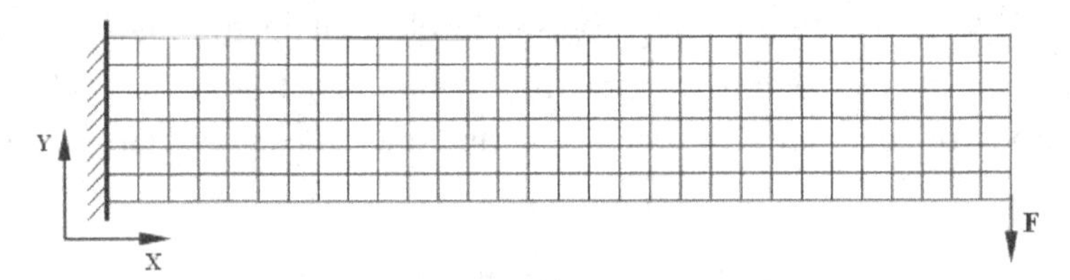

Fig. 2.10. Macroscopic FE model and its boundary conditions

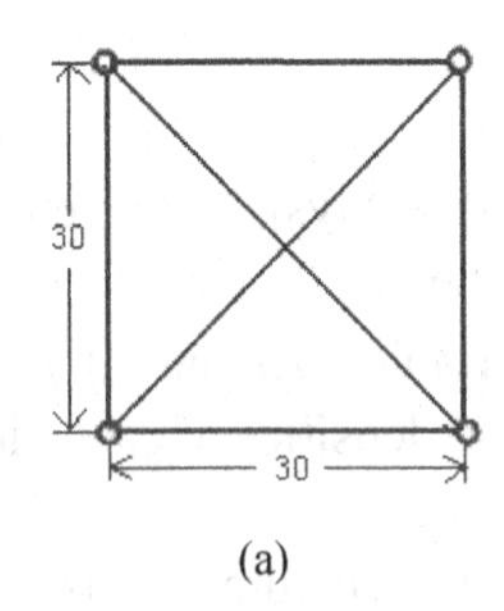

(a)

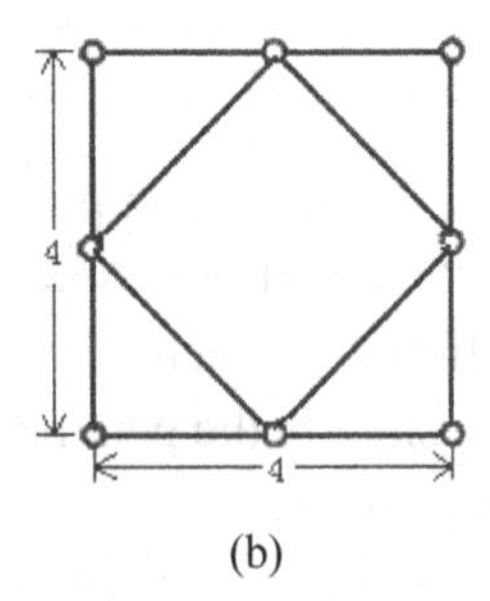

(b)

Fig. 2.11. Configurations of the micro truss sub-unit cells

As shown in Fig. 2.10 a cantilever structure which is composed of $n_x \times n_y = 30 \times 6$ macro four-node element, which is periodic truss unit cell, where n_x and n_y denote the number of macro element in the X-

and Y-directions, respectively. The left side of the structure is fixed in the two directions, and a downward concentrated load, which equals to 1000 is applied on the lower endpoint of the right side. Fig. 2.11 gives two configurations of basic sub-unit cell. Then n^2 basic sub-unit cells are utilized to construct the microstructure of one macro four-node element, where n is called the size factor and $n = [1,2,3,4]$ will be considered in the example. Notably, the extension of the unit cell only increases the number of trusses and the cross-sectional area of each truss is independent to others. In order to ensure the same material volume, the initial values about the length and cross-sectional area of the rods within the truss unit cell become $1/n$ of those in sub-unit cell shown in Fig. 2.11. Here the parameters $\overline{\varsigma^{\mathrm{MI}}}$ and $\bar{\varsigma}$ are set as 0.5 and 0.3 respectively, and V^{MI} is taken as a specified value 494.5.

Table 2.5 and Table 2.6 show the influence of the size effect on the optimized results under the linear and periodic boundary conditions, respectively. Linear or periodic boundary conditions imposed on the truss unit cell shown in Fig. 2.11(a) are equivalent when scale factor is taken as 1. Therefore, optimal results are also the same in the case. The objective functions under the two boundary conditions present the increasing tendency with the increase of scale factor n except for $n = 1$, which reflects size effect of the truss unit cell on the optimal results.

The reason for the phenomenon is that the increasing of the number of rods within the truss unit cell results in the decreasing of material distributed in the main load-bearing rods on the upper and lower horizontal boundaries. This means that more base material distributed in the internal rods of the truss unit cell with lower load carrying efficiency. However, the objective function with scale factor $n = 2$ is smaller than that with scale factor $n = 1$. An increasing of the number of rods within the truss unit cell leads to the decreasing of material distribution in the main load-bearing rods. Thus, the distribution of material is more efficient in the case. Therefore, balance should be found in the above contradiction to achieve the optimal distribution of base material.

Observing the material distribution of the lattice materials in the Table 2.5 and Table 2.6, we can find that base material is mainly distributed in the upper and lower horizontal rods of the truss unit cell to bear the stress caused by the concentrated force. The diagonal rods with relatively smaller cross-sectional area resist the shear force. And the vertical rods have the cross-sectional area approaching to the specified lower bound which realizes the topology optimization of the truss unit cell equivalently. Moreover, cross-sectional areas of the rods located on the lower boundary are larger than those situated on the upper boundary, and the cross-sectional areas of horizontal rods are much larger than those of diagonal rods. Obviously, it is consistent with the stress distribution characteristics of the cantilever structure, which demonstrates the feasibility and correctness of the two-scale concurrent optimization in some extents.

Table 2.5. Optimized results with different n under linear boundary conditions

n	Compliance	Macro topology	Material microstructure
1	3993		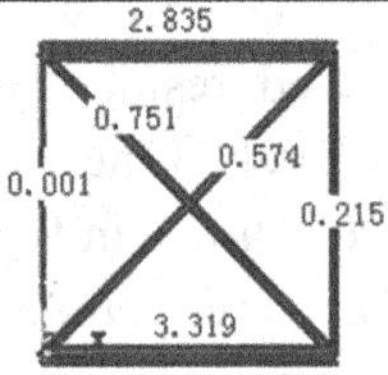
2	3849		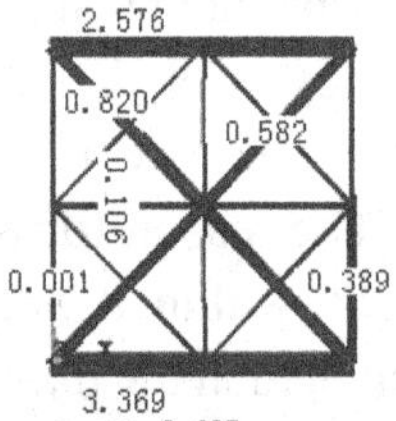
3	3981		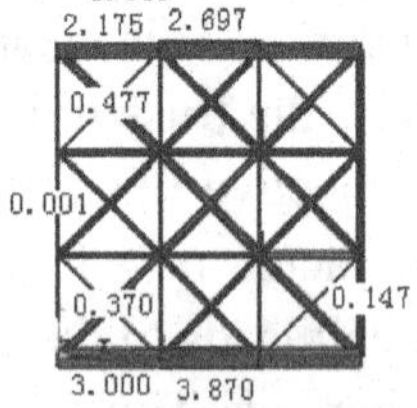
4	4002	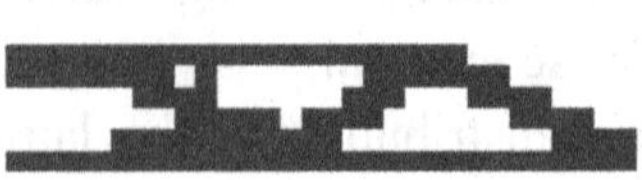	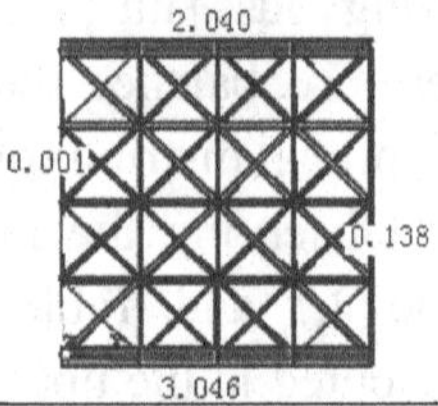

Table 2.6. Optimized results with different n under periodic boundary conditions

n	Compliance	macro topology	Material microstructure
1	3993		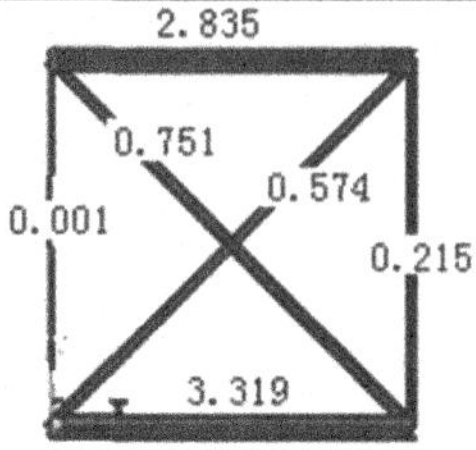
2	3873		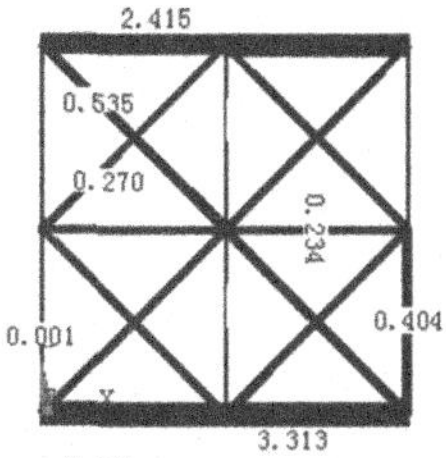
3	3995		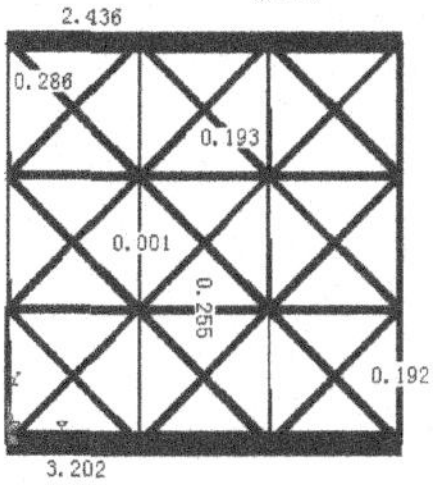
4	4033		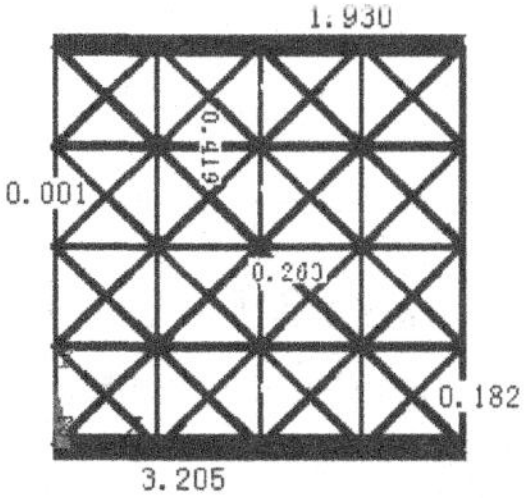

(2) **Example 2:** A discussion of parameters $\bar{\varsigma}$

The influence of parameter $\bar{\varsigma}$ (base material volume fraction) on the optimized results is considered under linear and periodic boundary conditions. A cantilever structure illustrated in Fig. 2.10 is still considered and the configuration of truss unit cell is shown in Fig. 2.11(b). In the optimization, the value of V^{MI} is taken as 27.3.

Table 2.7. Optimized results for varying $\bar{\varsigma}$ under linear bondary conditions with $\overline{\varsigma^{MI}}=0.4$

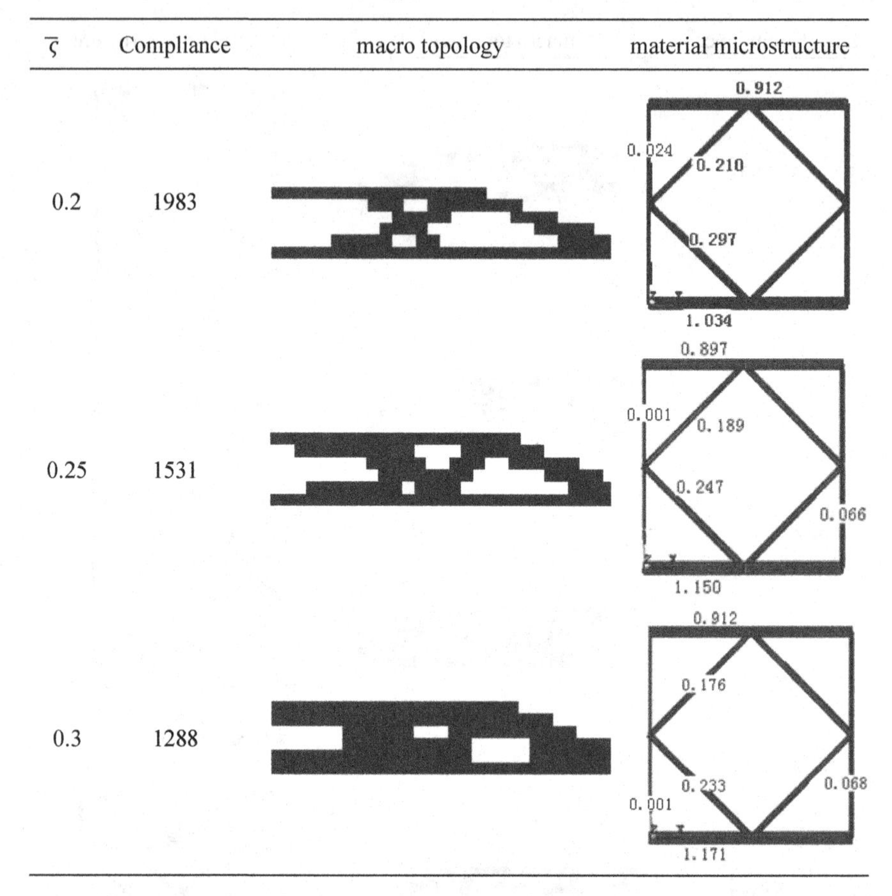

$\bar{\varsigma}$	Compliance	macro topology	material microstructure
0.2	1983		
0.25	1531		
0.3	1288		

Table 2.7 and Table 2.8, respectively, give the optimized results for varying $\bar{\varsigma}$ under linear and periodic boundary conditions with $\overline{\varsigma^{MI}} = 0.4$. It can be found that the corresponding objective functions under the two boundary conditions are almost equal at the same value of the $\bar{\varsigma}$. However, the macro topology of the structure and cross-sectional area of the rods within the truss unit cell are different obviously. Structural compliance under the two boundary conditions decreases with the increase of base material volume. In the microscale, the material distribution within the truss unit cell also follows the rules found in numerical example 1 and vary with the change of $\bar{\varsigma}$. In order to validate the optimization results

quantitatively, we model three designs for the case with the base material volume fraction $\overline{\varsigma^{MI}} = 0.25$ under periodic boundary conditions, that is, optimization results of the concurrent optimal design (i.e., the second row in Table 2.8), the design of material microstructure singly and the uniform configuration without optimization by the Direct Numerical Simulation (DNS) based on the optimized macro and micro topologies. The deformation results of the three design are listed in Appendix B, which can show the superiority of the concurrent optimization model and algorithm quantitatively in some extends.

Table 2.8. Optimized results for varying $\bar{\varsigma}$ under periodic boundary conditions with $\overline{\varsigma^{MI}} = 0.4$

$\bar{\varsigma}$	compliance	macro topology	material microstructure
0.2	1994		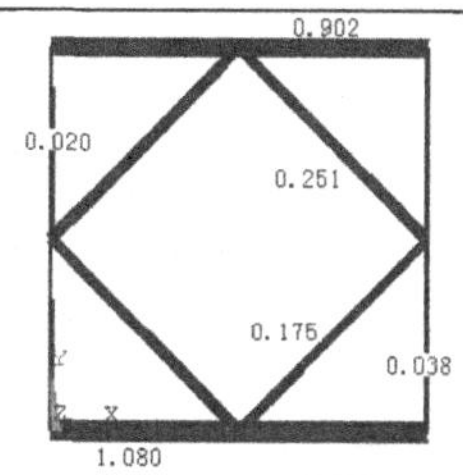
0.25	1547		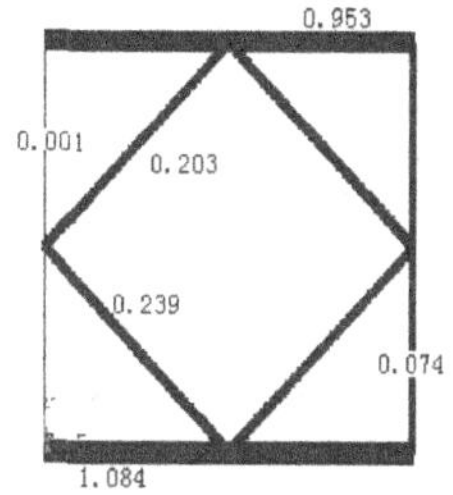
0.3	1293		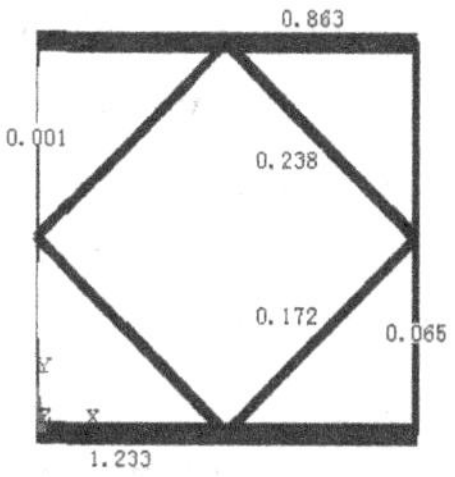

(3) Example 3: Validation of the superiority of two-scale concurrent optimization design

As shown in Fig. 2.12, an L-shaped beam composed of truss unit cell displayed in Fig. 2.11(b) is considered. The upper edge is fixed in the two axis directions, and a group of loads is applied along the right edge to simulate a moment. The parameters V^{MI}, $\overline{\varsigma^{\mathrm{MI}}}$ and $\bar{\varsigma}$ in the example are taken as 27.3, 0.5 and 0.3, respectively. The L-shaped beam can be divided into a, b, and c three parts as shown in Fig. 2.12. Both part a and part b whose side length is H are square. Side length of part c is L that is a variable determined by parameter n which is defined to adjust the weight between the vertical and horizontal parts and its values are taken as 0, 1, 3 and 5, respectively in the example.

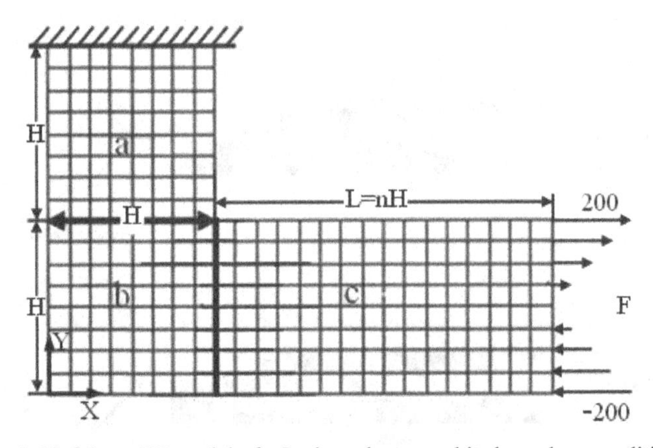

Fig. 2.12. Macro FE model of a L-shape beam and its boundary conditions

According to the loading characteristics of the various parts of the L-shape beam, we think that the material of "part a" should be located on the vertical rods of left and right bounds to obtain a larger moment of inertia in the Y-direction to bear the applied moment. The material of "part c" should be mainly distributed in the horizontal rods of upper and lower bounds to acquire a larger moment of inertia in X-direction to resist the external moment. However, due to the homogeneity of the truss unit cell in macro-scale, the above predictions are an obvious contradiction. In order to solve the contradiction, material distribution in truss unit cell should take into account the stress ratio of the two parts a and c, which relates to the choice of factor n between the two parts. The followings are the concurrent optimization results on the two-scales (C-I) with the change of factor n, which are compared with the optimized designs only for the truss unit cell in the microscale (C-S). The symbol 'C' above means the structural 'Compliance'.

By comparing Table 2.9 and Table 2.10, optimized results under the two boundary conditions are almost equal for the same factor n. And we can find due to the expansion of the design space, the distribution of material is more effective for the concurrent optimization on the two-scales. Therefore, objective functions of two-scale concurrent optimization are much smaller than those of optimized design only for the truss unit cell.

From Table 2.9 and Table 2.10, macro configurations and material microstructures under the two boundary conditions vary with the change of factor n. When parameter $n = 0$, L-shaped beam becomes a vertical cantilever beam, and material is distributed on the left and right boundaries of the macrostructure and correspondingly the left and right rods of the truss unit cell to resist the external moment. As the factor n increases, the weight of part c is increased and becomes the dominance gradually in the entire structure. Because part c mainly produces stress along the horizontal direction, more and more material is distributed in horizontal rods within the truss unit cell. As a result, the cross-sectional areas of horizontal rods are significantly greater than those of the vertical rods. Due to the shear stress in part b, a small part of the material is distributed in the diagonal rods within the truss unit cell. The topology configuration of part b shows base material mainly distributed in upright position in the macroscale, which is consistent with the feature of material distribution in the microscale that the cross-sectional areas of rods located on upper and right boundaries are obviously larger than those of rods located on lower and left boundaries. With the increase of parameter n, the cross-sectional areas of corresponding boundary rods approach gradually. Optimized results presented in Table 2.9 and Table 2.10 are consistent with the above predictions, which further validate the rationality and effectiveness of the two-scale concurrent optimized design.

Table 2.9. Comparison of optimized results of the two-scale and the one scale of the L-shaped beam under linear boundary conditions

n	C-S	C-I	Reduction	Macro structural topology	Material Microstructure
0	51.9	41.8	19.3%		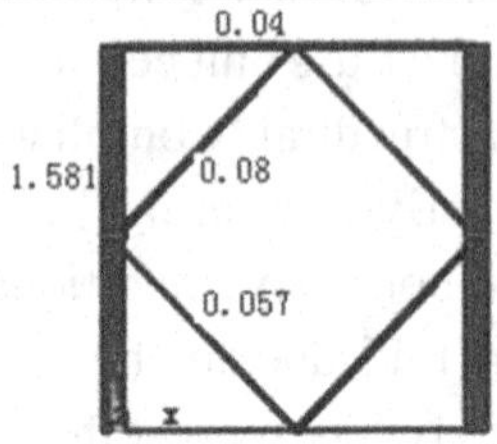
1	193.7	148.8	23.2%		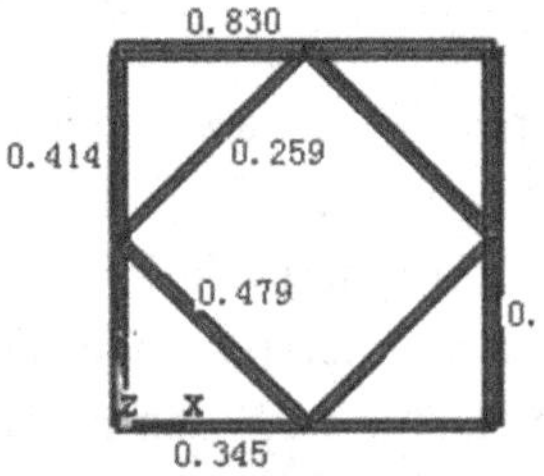
3	303.4	214.2	29.4%		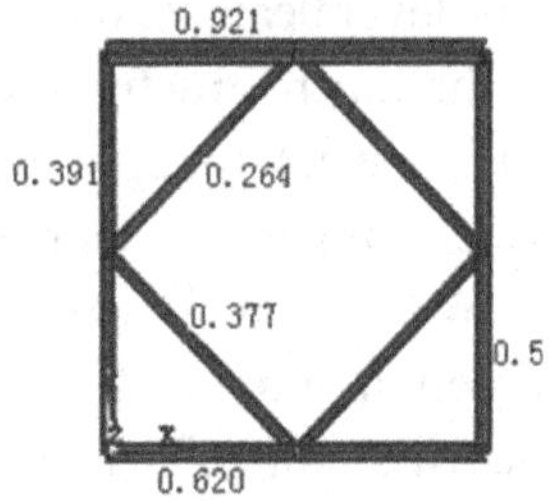
5	396.2	271.0	31.6%		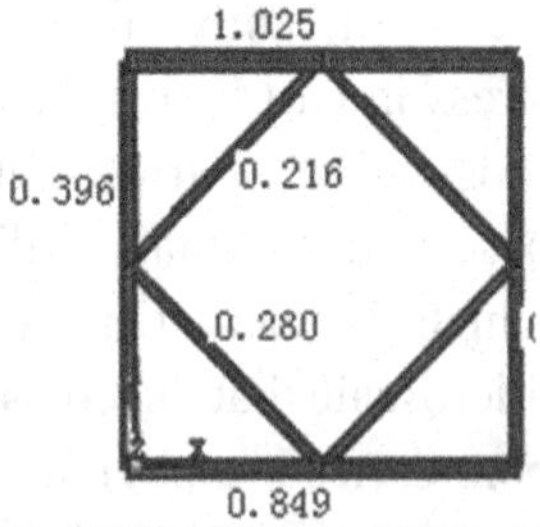

C-I: the concurrent optimization results on the two-scales with the change of parameter n
C-S: the optimized designs only for the truss unit cell in micro-scale with the change of parameter n

Table 2.10. Comparison of optimized results of two-scale and one scale of the L-shaped beam under periodic boundary conditions

n	C-S	C-I	Reduction	Macro structural topology	Material Microstructure
0	51.9	41.8	19.3%		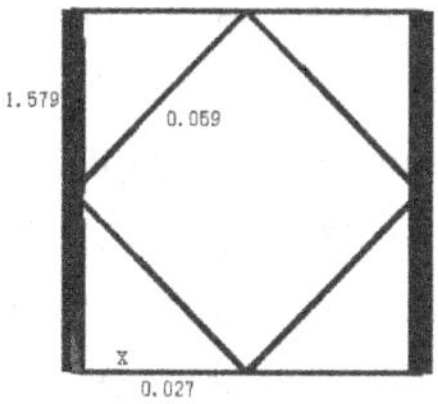 1.579 0.059 X 0.027
1	193.7	148.7	23.2%		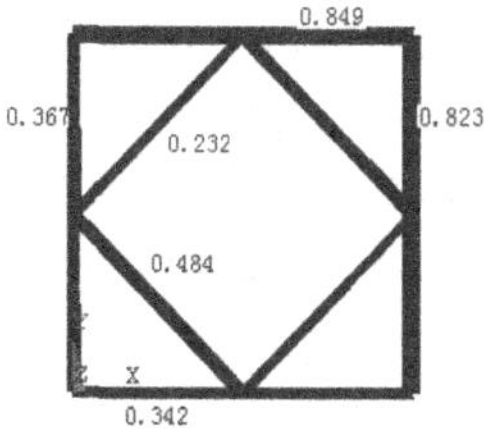 0.849 0.367 0.823 0.232 0.484 X 0.342
3	303.8	213.8	29.6%		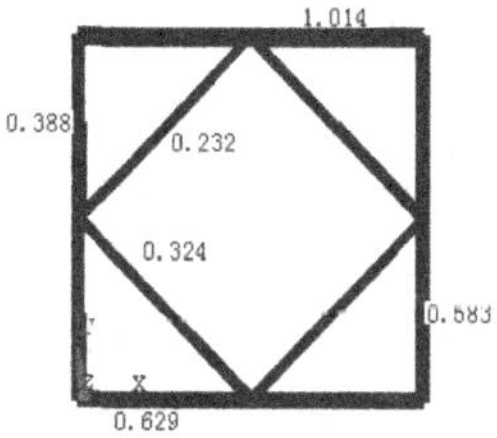 1.014 0.388 0.232 0.324 0.583 X 0.629
5	396.2	271.9	31.4%		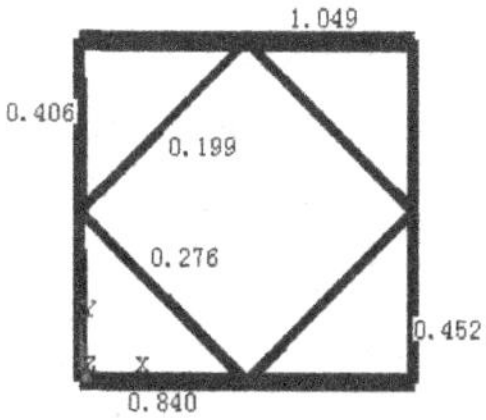 1.049 0.406 0.199 0.276 0.452 X 0.840

2.3 *Concluding remarks*

The PAMP formulation presented in this chapter provides a general framework for concurrent structural and material topology optimization. In PAMP formulation, structured material with microstructure to be designed is explicitely introduced in the macro-structure. Based on that, topology descriptions and design variables for macro-structure and microstructure are allowed to define independently. The effective property of structured material is the only coupling between the macro and micro scales. For example, in Section 2.1 both the macro-structural topology and material microstructural topology are described by the density approach and are coupled by AH method. And in Section 2.2 the structural topology is described by density distribution and the microstructural topology is described by truss structures and are coupled by EMsFEM. The independence of macro and micro design variables also makes the sensitivity and optimization less computationally expensive. For comparison, the approach in literatures which uses the density distribution in microstructure to describe the macro-structural topology introduces complicated coupling between the two-scale.

With the formulation PAMP, this chapter develops two concurrent topology design methods based on AH method and EMsFEM, respectively. The designs are aimed at searching the optimum macro-structure composed of an optimum porous anisotropic material. Design variables in two-scales are independently defined and the microstructure is no longer restricted to certain specific configurations, which makes it possible to take advantage of various porous or truss-like materials.

(1) In the concurrent design based on AH method, the PAMP scheme is proposed to penalize porous anisotropic material in the macro-scale design, and the conventional SIMP scheme is used in the microscale. Designs in both scales are integrated into one optimization problem and solved concurrently. Constraint parameters $\overline{\varsigma^{\mathrm{MI}}}$, $\bar{\varsigma}$ and $\bar{\gamma}$ are numerically discussed. Some innovative configurations of macro and microstructures are presented.

(2) In the concurrent design based on EMsFEM, comparisons between the linear and periodic boundary conditions are first performed. It is found that the different boundary conditions have significant effect on the optimized results. Meanwhile, the design factor $\bar{\varsigma}$ and the number of extensions of truss unit cell are studied as well. After that, the superiority of the concurrent multiscale optimization relative to the single-scale design is demonstrated through numerical examples. Moreover, validations of optimized results using DNS show that the periodic boundary conditions can better capture the structural response.

We strongly underline the uniformity of material microstructure in the macroscale. This assumption will inevitably weaken the benefit derived from the optimum design, but could result in an easier manufacturing process and then illuminate practical applications of the results. It will be an interesting and challenging work to extend this concurrent optimization method to multifunctional application fields (e.g., heat transfer, vibration isolation, and mechanical requirements) and 3D problems.

It should be mentioned that, inspired by the framework of the PAMP approach, some research works have been presented with considering non-uniform or partitioned microstructure. Meanwhile, that will be also very challenging and interesting to implement macrostructure topology by Level Set (LS) or Moving Morphable Component (MMC) schemes considered stress, frequency, and fatigue constraints.

It is worth emphasizing that the motivation to develop the concurrent multiscale optimization based on EMsFEM is for the consideration of size effect, which is omitted in the AH method due to the infinitely small assumption of the microstructural size. The size related analysis and optimization will be further discussed in Chapter 3. However, the theory of AH has a strict mathematical deducing. It can be well adapted in the structural analysis and optimization when the size effect is not obvious.

Chapter 3

Statics: Strength Optimization and Size Effects

The structural compliance studied in Chapter 2 is commonly aimed at the global behavior of a structure, and it can be regarded as a representation of the average stress over the structure, in some situations. However, the failures caused by the structural stress usually exhibit obvious local characteristics, e.g., the stress concentrations. Therefore, to improve the securities of a structure, the optimization for the structural strength is needed but challenging as well [113].

Considering the possibility of high stress gradient in strength minimization problem, in the first part of this chapter we model the lattice material as micropolar continua representation. Two classes of design variables, i.e., the relative density and the cell size distribution of truss-like materials, are to be determined by optimization under a given total material volume constraint. Three different formulations are introduced to define the stress-based optimization problem, i.e., minimization of the maximum stress at a given point, minimizing the maximum stress in specified points set, and maximizing the strength reserve at the critical area. Then, differences among the three objectives are revealed through numerical examples.

In the second part of this chapter, lightweight design under stress constraints based on EMsFEM is studied. Before optimization, the size effect of lattice structure is investigated. Then, some suggestions based on numerical experience are given to illustrated the appropriate applications of AH method and EMsFEM. Based on the understanding of the size effect, the EMsFEM-based optimization for lightweight design with stress constraints is established subsequently.

The work of this chapter is mainly related with references [114, 115].

3.1 *Stress optimization of lattice materials*

3.1.1 *Representation of 2D periodic lattice materials as micropolar continua*

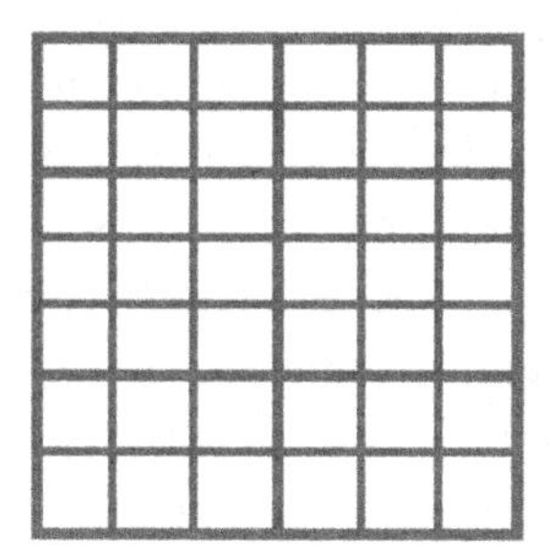

Fig. 3.1. Illustration of the configuration of a 2D lattice structure

The illustration of the configuration of a 2D lattice structure is shown in Fig. 3.1. For simplicity, 'Lattice structure' and 'Lattice material' refers specifically to the macro and micro structure, respectively. For a straightforward analysis approach, the overall properties of lattice structure are evaluated via a discrete modeling where the individual cell walls of lattice material are modeled using beams or rods. The detailed discrete model above gives very accurate stresses and strains in cell walls. However, an expensively computational cost is needed, which is unacceptable for optimization of practical lattice structures. The equivalence approach, which model lattice material as a continuum can significantly reduce the amount of cox`mputation. Thus, it provides a solid basis for investigation of the optimum distribution of lattice material if the equivalence guarantees the precision of analysis. There is a large number of literatures on this type of equivalence, among which the Representative Volume Element (RVE) method and AH method [9, 116] are most representative. In most of these literatures, lattice materials are modeled as a classical continuum where it assumes only two translational degrees of freedom for each material point and the interaction between neighboring material points is realized by stress tractions.

As pointed out by Fleck and Hutchinson [117] and Adachi et al. [118] for those cases in which the gradient of the strain is large, the classical continuum theory is not accurate enough. To capture this non-locality, independent rotational degrees of freedom and couple stresses in addition to the usual Cauchy-type stress at each material point have been introduced to enrich the classical theory. This theory is referred to as micro polar theory [119, 120]. Since the solid part of lattice material is distributed discretely in space, and there is a large gradient of strain around the hole, we need to model the lattice material as a micropolar continuum to guarantee enough computation precision. It is worth pointing out that the approach presented here only evaluates equivalent stresses in the structures composed of lattice material, the effects of maximum stress inside a unit cell may not be analyzed exactly. To consider the effects of maximum stress inside a unit cell of composites, readers are recommended to refer to the work by Lipton [121, 122].

3.1.2 *Governing equations of linear micropolar elasticity*

It gives the fundamental equations in linear micropolar elasticity theory [119] below. The balance of linear momentum:

$$\sigma_{ji,j} + f_i = 0 \tag{3.1}$$

balance of moment of momentum:

$$e_{ijk}\sigma_{jk} + m_{ji,j} + g_i = 0 \tag{3.2}$$

the kinematic relations:

$$\varepsilon_{ji} = u_{i,j} - e_{kji}\phi_k \tag{3.3}$$

$$\kappa_{ji} = \phi_{i,j} \tag{3.4}$$

the constitutive equations:

$$\sigma_{ji} = A_{jikl}\varepsilon_{kl} + B_{jikl}\kappa_{kl} \tag{3.5}$$

$$m_{ji} = B_{klji}\varepsilon_{kl} + C_{jikl}\kappa_{kl} \tag{3.6}$$

These are supplemented by displacement and micro-rotation boundary conditions,

$$u_i = u_i^P \tag{3.7}$$

$$\phi_i = \phi_i^P \tag{3.8}$$

or traction boundary conditions

$$\sigma_{ji} n_j = t_i^P \tag{3.9}$$

$$m_{ji} n_j = c_i^P \tag{3.10}$$

In the above equations, $\boldsymbol{\sigma}$ is the non-symmetric stress tensor, $\boldsymbol{m}$ is the couple stress tensor, $\boldsymbol{f}$ is the applied body force vector, $\boldsymbol{g}$ is the applied body moment vector, $\boldsymbol{\varepsilon}$ is the strain tensor, $\boldsymbol{\phi}$ is the micro-rotation vector, and $\boldsymbol{\kappa}$ is the curvature tensor. The prescribed values of displacement, microrotation, stress traction and couple stress traction are denoted by $\boldsymbol{u}^{\mathrm{p}}$, $\boldsymbol{\phi}^{\mathrm{p}}$, $\boldsymbol{t}^{\mathrm{p}}$, $\boldsymbol{c}^{\mathrm{p}}$, respectively. $\boldsymbol{n}$ is an outward unit normal vector defined on the boundary. The constitutive behavior is described by fourth-rank tensors A, B, and C. Since the periodic cell structures considered here are centrally symmetric, the components of the coupling term in the constitutive description $\boldsymbol{B}$ are identically zero.

3.1.3 *Micropolar constitutive relations for structures with square unit cells*

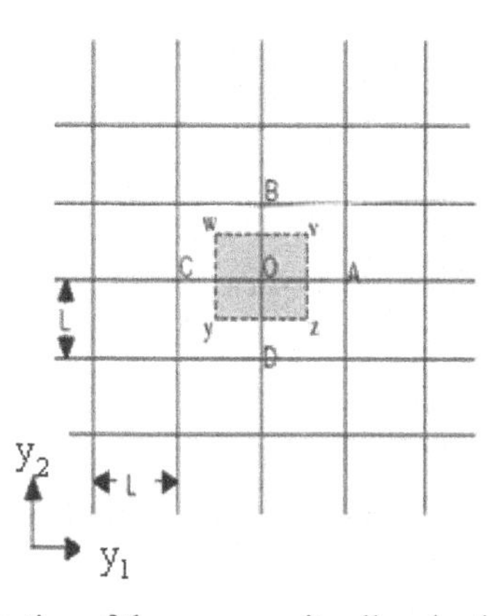

Fig. 3.2. Illustration of the square unit cells selection of a lattice structure

Fig. 3.2 is a illustration figure of a lattice structure with square unit cells. The micropolar constitutive equations for this kind of lattice material can be derived from the concept of strain energy equivalence through simplifying individual members of the unit cell as beams by Kumar and McDowell [123]. And the derivations are restated here briefly.

The strain energy of the square unit cell is due to the deformation of members OA, OB, OC, and OD. Let displacement and micro-rotation at joint O be $\boldsymbol{U}$, and ϕ respectively, the strain energy density is given by

$$\begin{aligned} w &= \frac{1}{2A^R}(W^{OA} + W^{OC} + W^{OB} + W^{OD}) \\ &= \frac{1}{2L^2}\begin{pmatrix} E'L^2U_{1,1}^2 + E'L^2U_{2,2}^2 + 12k(U_{2,1} - \phi)^2 + \\ 12k(U_{1,2} + \phi)^2 + 4k(L\phi_{,1})^2 + 4k(L\phi_{,2})^2 \end{pmatrix} \end{aligned} \tag{3.11}$$

where $A^{\mathrm{R}} = L^2$ is the area of the unit cell "w-v-z-y". In Eq. (3.11), $E' = \mathrm{E}h/L$, $k = \mathrm{E}h^3/(12L)$, E is Young's modulus of the cell wall material, h is the thickness of the cell wall, and L is the size of cell structure. Thus, stresses and couple stresses can be derived from strain energy density function, as shown below in Eq. (3.12).

$$\begin{aligned} \sigma_{11} &= \frac{\partial w}{\partial \varepsilon_{11}} = \frac{\partial w}{\partial U_{1,1}} = E'\varepsilon_{11} \\ \sigma_{22} &= \frac{\partial w}{\partial \varepsilon_{22}} = \frac{\partial w}{\partial U_{2,2}} = E'\varepsilon_{22} \\ \sigma_{12} &= \frac{\partial w}{\partial \varepsilon_{12}} = \frac{\partial w}{\partial (U_{2,1} - \phi)} = \frac{12k}{L^2}\varepsilon^{12} \\ \sigma_{21} &= \frac{\partial w}{\partial \varepsilon_{21}} = \frac{\partial w}{\partial (U_{1,2} + \phi)} = \frac{12k}{L^2}\varepsilon^{21} \\ m_{13} &= \frac{\partial w}{\kappa_{13}} = \frac{\partial w}{\partial \phi_{,1}} = 4k\kappa_{13} \\ m_{23} &= \frac{\partial w}{\kappa_{23}} = \frac{\partial w}{\partial \phi_{,2}} = 4k\kappa_{23} \end{aligned} \tag{3.12}$$

Based on the topology and dimension of unit cells shown in Fig. 3.2, we can obtain:

$$\begin{aligned} \rho^* &= \frac{L^2 - (L-h)^2}{L^2}\rho_s \Rightarrow \rho = \frac{\rho^*}{\rho_s} = 2\frac{h}{L}\left(1 - \frac{1}{2}\frac{h}{L}\right) \\ &\Rightarrow \frac{h}{L} = 1 - \sqrt{1-\rho} \end{aligned} \tag{3.13}$$

where ρ_s is the density of cell wall material, ρ^* is the density of the lattice material, ρ is the relative density of lattice material. Substitute Eq. (3.13) into Eq. (3.12), the constitutive equations can be rewritten as:

$$\begin{aligned} &\sigma_{11} = E\left(1-\sqrt{1-\rho}\right)\varepsilon_{11} && \sigma_{22} = E\left(1-\sqrt{1-\rho}\right)\varepsilon_{22} \\ &\sigma_{12} = E(1-\sqrt{1-\rho})^3\varepsilon_{12} && \sigma_{21} = E(1-\sqrt{1-\rho})^3\varepsilon_{21} \\ &m_{13} = \frac{1}{3}E(1-\sqrt{1-\rho})^3L^2\kappa_{13} && m_{23} = \frac{1}{3}E(1-\sqrt{1-\rho})^3L^2\kappa_{23} \end{aligned} \tag{3.14}$$

3.1.4 *Finite element implementation for micropolar elasticity*

The principle of virtual work for micropolar elasticity (assuming zero body forces and moments) is given by

$$\begin{aligned} &\int_{\Omega}\left(\sigma_{ij}\delta u_{j,i} - e_{ijk}\sigma_{ij}\delta\phi_k + m_{ij}\delta\phi_{j,i}\right)dV \\ &= \int_{\partial\Omega}(T_i\delta u_i + Q_i\delta\phi_i)dA \end{aligned} \tag{3.15}$$

where $\delta\boldsymbol{u}$, $\delta\boldsymbol{\phi}$ are the virtual displacement vector and the virtual micro-rotation vector, respectively. $\boldsymbol{T}$, $\boldsymbol{Q}$ are applied surface traction and surface moment vectors, respectively.

Here planar eight-node isoperimetric elements are developed with the micropolar theory. In this element, each node has three degree of freedoms—two in-plane displacement components u_1 and u_2, and an in-plane micro-rotation component ϕ_3. The element stiffness matrix can be expressed as:

$$\boldsymbol{k}_e = \int_{\Omega^e}\boldsymbol{B}^T\,\boldsymbol{D}\,\boldsymbol{B}dV \tag{3.16}$$

where $\boldsymbol{D}$ is the constitutive matrix from Eq. (3.14), and $\boldsymbol{B}$ is the strain-displacement matrix shown below.

$$\boldsymbol{B}_i = \begin{bmatrix} \frac{\partial N_i}{\partial x_1} & 0 & 0 & \frac{\partial N_i}{\partial x_2} & 0 & 0 \\ 0 & \frac{\partial N_i}{\partial x_2} & \frac{\partial N_i}{\partial x_1} & 0 & 0 & 0 \\ 0 & 0 & -N_i & N_i & \frac{\partial N_i}{\partial x_1} & \frac{\partial N_i}{\partial x_2} \end{bmatrix}^T_{i=1-8} \tag{3.17}$$

where N_i are the shape functions and their detailed formulations, and the global system of algebraic equations assemblage can be found in reference. A computer program was developed in FORTRAN90 to implement the finite element formulation. The benchmark problem of stress concentration around a circular notch in a square plate made of homogeneous orthotropic micropolar solid in reference [124] was solved.

3.1.5 *Optimization formulations*

We choose three parameters to describe lattice structure: local relative density (on macro scale), unit cell size (on micro scale), and cell wall thickness (on micro scale). Since the three parameters satisfy Eq. (3.13) for a square unit cell, thus the number of independent parameters is reduced to two. Here we choose relative density and unit cell size as the independent design variables. The equivalent micropolar continuum of lattice structure is meshed into a number of FEM elements. The material distribution or the arrangements of cells is uniform in each element and varies from element to element. Our optimization aims at finding the optimum distribution of the cell density and cell size over the design domain to minimize the objective and satisfy the given constraint set.

(1) Three formulations of objective functions

The following three different objective functions are considered.

Model-1: To minimize the stress at a given point where stress concentration appears for the initial uniform design;

Model-2: To minimize the maximum stress in the specified points set which we care most. Minimization of stress at a given point may elevate stresses at other points. Then some points we care are chosen as a specified point set $\boldsymbol{V}_0$, in which the maximum stress should be minimized;

Model-3: To maximize the strength reserve of the part around the hole, or equivalently, to minimize the ratio of stress around the hole over the effective yield strength. Since the effective yield strength is dependent on the relative density of lattice materials, Model-3 differs from Model-1 generally. Based on reference [125], normalized effective yield strength σ_{pl} for lattice structural with square unit cells can be written as $\sigma_{\mathrm{pl}}/\sigma_{\mathrm{ys}} = 0.5\rho$, where σ_{ys} is the yield strength of solid cell wall material.

The mathematical formulation for Model I can be expressed as:

Model-1:

$$
\begin{aligned}
\text{find} \quad & \rho_i, L_i \\
\min \quad & \Psi = (\boldsymbol{\sigma}^C)^2 \\
\text{s.t.} \quad & \int_{\Omega} \rho_i d\Omega = S_0 \\
& L_i\left(1 - \sqrt{1-\rho_i}\right) \geq \underline{h} \\
& 0 \leq \underline{\rho} \leq \rho_i \leq 0.3; \underline{L} \leq L_i \leq \overline{L}
\end{aligned}
\tag{3.18}
$$

For Model-2 and Model-3, the objectives should be replaced by

Model-2:

$$
\min \quad \Psi = (\underset{i \in V_0}{Max}\, \boldsymbol{\sigma}^i)^2 \tag{3.19}
$$

and

Model-3:

$$
\min \quad \Psi = (\boldsymbol{\sigma}^C)^2/(0.5\rho_C) \tag{3.20}
$$

respectively. Model-2 and Model-3 have the same constraints as those of Model-1.

$\boldsymbol{\sigma}^{\mathrm{C}}$ and $\boldsymbol{\sigma}^{\mathrm{i}}$ in Eqs. (3.18)-(3.20) are some stress measure or some component of nodal stress vector. $\boldsymbol{\sigma}^{\mathrm{C}}$ and ρ_{C} denote the stresses and relative density of lattice material at points with stress concentration, such as stress at point P shown in Fig. 3.4 and Fig. 3.9. $\boldsymbol{V}_0$ is the set of specified points, which we care most. S_0 is the prescribed material volume of structures. The cell wall thickness h is subjected to a lower bound $\underline{h} = 0.1$mm, which is determined by the limitation of manufacture. The two design variables ρ and L have the upper and lower bounds due to manufacture considerations [126, 127].

(2) Constraints transformation

The constraints in Eq. (3.18) are transformed into a standard form as follows

$$\begin{aligned} g_j(\boldsymbol{X}) &\le 0 \qquad (j=1,2) \\ \boldsymbol{X}_i^L \le \boldsymbol{X}_i &\le \boldsymbol{X}_i^U \quad (i=1,2) \end{aligned} \tag{3.21}$$

where, $g_1(\boldsymbol{X}) = \boldsymbol{s}^{\mathrm{T}}\boldsymbol{\rho}/S_0 - 1$, $g_2(\boldsymbol{X}) = 1 - L(1-\sqrt{1-\rho})/\underline{h}$, $\boldsymbol{X}_1 = \boldsymbol{\rho}$, $\boldsymbol{X}_2 = \boldsymbol{L}$, and their lower and upper bounds are specified as $\boldsymbol{X}_1^{\mathrm{L}} = \underline{\rho} = 0.01$; $\boldsymbol{X}_1^{\mathrm{U}} = \overline{\rho} = 0.3$ and $\boldsymbol{X}_2^{\mathrm{L}} = \underline{L} = 0.1$ mm, $\boldsymbol{X}_2^{\mathrm{U}} = \overline{L} = 15$mm. Boldface letters $\boldsymbol{s}$, $\boldsymbol{\rho}$ and $\boldsymbol{L}$ denote the vectors composed of area, relative density and cell size of each element respectively.

3.1.6 *Sensitivity analysis*

(1) Sensitivities of objective functions

Based on the formula of nodal stress vector $\boldsymbol{\sigma} = \boldsymbol{DBu}$, we can obtain sensitivities of $\boldsymbol{\sigma}$ at a nodal point by the direct method

$$\frac{\partial \boldsymbol{\sigma}}{\partial \rho_i} = \boldsymbol{B}^T \frac{\partial \boldsymbol{D}}{\partial \rho_i}\boldsymbol{u} + \boldsymbol{D}\frac{\partial \boldsymbol{u}}{\partial \rho_i}, \qquad \frac{\partial \boldsymbol{\sigma}}{\partial L_i} = \boldsymbol{B}^T \frac{\partial \boldsymbol{D}}{\partial L_i}\boldsymbol{u} + \boldsymbol{B}^T\boldsymbol{D}\frac{\partial \boldsymbol{u}}{\partial L_i} \tag{3.22}$$

$\boldsymbol{u}$ denotes the nodal displacement vector. Nodal stress vector $\boldsymbol{\sigma}$ can be expressed as a simple function in terms of $\boldsymbol{D}$ and $\boldsymbol{u}$ and its sensitivities can be easily obtained based on Eq. (3.22),where $\partial\boldsymbol{D}/\partial\rho_i$ and $\partial\boldsymbol{D}/\partial L_i$ can be derived from Eq. (3.14), $\partial\boldsymbol{u}/\partial\rho_i$ and $\partial\boldsymbol{u}/\partial L_i$ can be obtained by solving the two equations below

$$\boldsymbol{K}\frac{\partial \boldsymbol{u}}{\partial \rho_i} = -\frac{\partial \boldsymbol{K}}{\partial \rho_i}\boldsymbol{u}, \qquad \boldsymbol{K}\frac{\partial \boldsymbol{u}}{\partial L_i} = -\frac{\partial \boldsymbol{K}}{\partial L_i}\boldsymbol{u} \tag{3.23}$$

$$\frac{\partial \boldsymbol{K}}{\partial \rho_i} = \frac{\partial \boldsymbol{K}_i^e}{\partial \rho_i} = \frac{\partial \int_{V^e} (\boldsymbol{B}_i^e)^T \boldsymbol{D}_i^e \boldsymbol{B}_i^e dV^e}{\partial \rho_i} = \int_{V^e} (\boldsymbol{B}_i^e)^T \frac{\boldsymbol{D}_i^e}{\partial \rho_i} \boldsymbol{B}_i^e dV^e \tag{3.24}$$

$$\frac{\partial \boldsymbol{K}}{\partial L_i} = \frac{\partial \boldsymbol{K}_i^e}{\partial L_i} = \frac{\partial \int_{V^e} (\boldsymbol{B}_i^e)^T \boldsymbol{D}_i^e \boldsymbol{B}_i^e dV^e}{\partial L_i} = \int_{V^e} (\boldsymbol{B}_i^e)^T \frac{\boldsymbol{D}_i^e}{\partial L_i} \boldsymbol{B}_i^e dV^e \tag{3.25}$$

So, for Model-1, we have the sensitivities of the objective function

$$\frac{\partial \Psi}{\partial \rho_i} = 2(\boldsymbol{\sigma}^C)\frac{\partial \boldsymbol{\sigma}^C}{\partial \rho_i}, \qquad \frac{\partial \Psi}{\partial L_i} = 2(\boldsymbol{\sigma}^C)\frac{\partial \boldsymbol{\sigma}^C}{\partial L_i} \tag{3.26}$$

For Model-2: Since the objective function is non-differentiable, we apply the p-norm method [128] to approximate Eq. (3.19) by Eq. (3.27). with which equivalence can be obtained when p tends to infinity.

$$min\ \widetilde{\Psi} = \left(\left(\sum_{i\in \boldsymbol{V}_0} (\boldsymbol{\sigma}^i)^p\right)^{\frac{1}{p}}\right)^2 \tag{3.27}$$

Accordingly, its sensitivities are given as follows.

$$\frac{\partial \widetilde{\Psi}}{\partial \rho_i} = 2\left(\left(\sum_{i\in \boldsymbol{V}_0} (\boldsymbol{\sigma}^i)^p\right)^{\frac{1}{p}}\right)\left(\sum_{i\in \boldsymbol{V}_0} (\boldsymbol{\sigma}^i)^p\right)^{\frac{1}{p}-1} \sum_{i\in \boldsymbol{V}_0} (\boldsymbol{\sigma}^i)^{p-1}\frac{\partial \boldsymbol{\sigma}^i}{\partial \rho_i} \tag{3.28}$$

$$\frac{\partial \widetilde{\Psi}}{\partial L_i} = 2\left(\left(\sum_{i\in \boldsymbol{V}_0} (\boldsymbol{\sigma}^i)^p\right)^{\frac{1}{p}}\right)\left(\sum_{i\in \boldsymbol{V}_0} (\boldsymbol{\sigma}^i)^p\right)^{\frac{1}{p}-1} \sum_{i\in \boldsymbol{V}_0} (\boldsymbol{\sigma}^i)^{p-1}\frac{\partial \boldsymbol{\sigma}^i}{\partial L_i} \tag{3.29}$$

For Model-3:

$$\frac{\partial \Psi}{\partial \rho_i} = \frac{4\boldsymbol{\sigma}^c}{\rho_c}\frac{\partial \boldsymbol{\sigma}^c}{\partial \rho_i} - 2\left(\frac{\boldsymbol{\sigma}^c}{\rho_c}\right)^2\frac{\partial \rho_c}{\partial \rho_i} \tag{3.30}$$

$$\frac{\partial \Psi}{\partial L_i} = \frac{4\boldsymbol{\sigma}^c}{\rho_c}\frac{\partial \boldsymbol{\sigma}^c}{\partial L_i} \tag{3.31}$$

where $\partial \boldsymbol{\sigma}/\partial \rho_i$ and $\partial \boldsymbol{\sigma}/\partial L_i$ can be obtained from Eqs. (3.22)-(3.24).

(2) Sensitivities of constraints

$$\frac{\partial g_1}{\partial \rho_i} = \frac{s_i}{S_0}, \qquad \frac{\partial g_1}{\partial L_i} = 0, \qquad (i = 1, \cdots, n) \tag{3.32}$$

$$\frac{\partial g_{2I}}{\partial \rho_i} = \frac{-L_i}{2\underline{h}}\frac{1}{\sqrt{1-\rho_i}}\delta_{Ii}, \tag{3.33}$$

$$\frac{\partial g_{2I}}{\partial L_i} = \frac{1}{h}\left(\sqrt{1-\rho_i} - 1\right)\delta_{Ii}(I, i = 1, \cdots, n)$$

where n is the number of elements.

3.1.7 *Numerical examples*

In examples below, the following parameters are prescribed: Young's module $\mathrm{E} = 1.0 \times 10^{11}\mathrm{Pa}$ for cell wall solid material of microstructure. The relative density and cell size of the initial design is uniform over the domain, $\rho = 0.1$ and cell size is $\mathrm{L} = 5\mathrm{mm}$. The prescribed material volume is one-tenth of that of design space all filled with solid materials. SQP algorithm [126] is used for all numerical examples.

(1) **Example 1:** Stress optimization around a hole in a circle plate

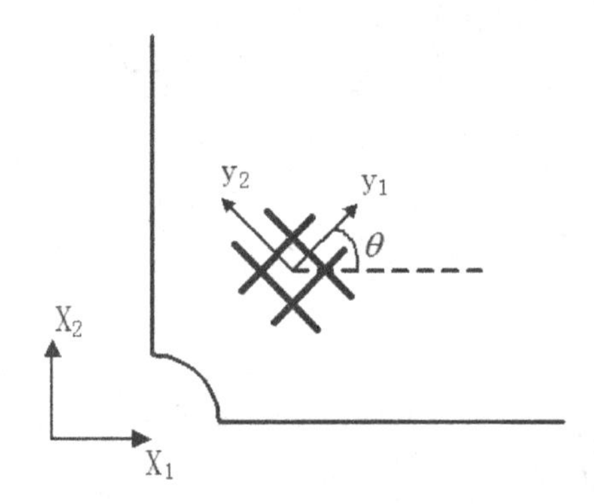

Fig. 3.3. Definition for coordinates and ply angle

First, we define the angle θ between the local orientation y_1-y_2 of the square unit cell and the global coordinates X_1-X_2 as material ply angle, as shown in Fig. 3.3. To meet the manufacturing requirements, the uniform ply angle is utilized over the structure.

A hollow circular plate of radius 400mm made of orthotropic lattice structure with the square unit cell is shown in Fig. 3.4. The outer edge is subjected to prescribed uniform pressure $p = 1.0$. First, it assumes that the ply angle of lattice material is fixed to zero, i.e., the local orientation y_1-y_2 of square unit cells coincides with the global X_1-X_2 directions, and then the difference among the objectives defined by Model 1, 2 and 3 are studied. The influence of ply angles on optimized results will be investigated. Since the material is not axisymmetric one, the structure is only symmetric with respect to the global coordinate axis X_1-X_2, a quarter of plate will be considered. On vertical edge, displacement in the X_1 direction and the micro-rotation ϕ are set to zero, whereas on horizontal edge, displacement in the X_2 direction and the micro-rotation ϕ are constrained. The edge of the inner hole is a traction free boundary. Other dimensions can be found in Fig. 3.4.

The maximum stress appears at point P in X_2 direction for the uniform initial design of homogeneous lattice structure. Since that, we choose $\boldsymbol{\sigma}^{C} = \boldsymbol{\sigma}_{X_2}^{P}$ as objective stress in Model-1. And the nodes on the bottom horizontal thicker solid line in Fig. 3.4 (whose X_2 coordinates are 0 while X_1 coordinates range from 20 to 400) are chosen as specified points set $\boldsymbol{V}_0$ in Model-2.

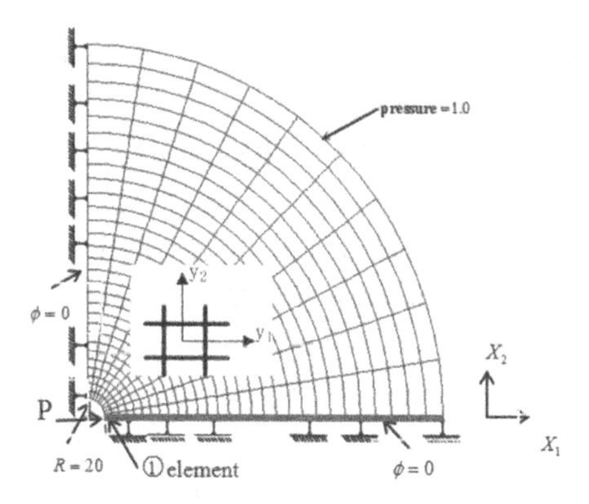

Fig. 3.4. Model of geometry and FEM for a quarter of the circular plate with a circular hole

For ease of comparison, without losing generality, we assume all elements on the same circumference have identical material properties. The results are summarized in Table 3.1 for the case of 45-element along radius. Mesh dependency is also investigated with three meshes composed of 30, 45 and 60 elements along radius, respectively and the detailed material distributions are shown in Fig. 3.5 to Fig. 3.7. $\sigma_{X_2}^{\text{Max}}$ is the maximum stress in X_2-direction of points set $\boldsymbol{V}_0$.

Table 3.1. Comparison of stress between uniform design and optimum design

Model	Initial objective function value	Objective function value after optimization	$\sigma_{X_2}^{Max}$
1	57.37 (stress around the hole 7.57)	0.34 (stress around the hole 0.58)	5.14
2	57.37 (stress around the hole 7.57)	6.36 (stress around the hole 2.52)	2.52
3	1147.41 (stress around the hole 7.57)	151.64 (stress around the hole 4.77)	4.77

Results in Table 3.1 show that for the uniform initial design the stress at the point P is 7.57, which leads to objective function values 57.37, 57.37 and 1147.41 for Model-1, Model-2 and Model-3, respectively. In Model-1, the stress at point P is minimized to 0.14. However, the maximum stress emerges at another point on the horizontal axis and reaches 5.14, which is certainly not desired. In Model-2, maximum stress on the horizontal axis can be reduced to 2.52. The further discussions on the numerical examples are given below.

a) For Model-1 and Model-2, the relative density has shown the following trends of distribution: low relative density in the immediate proximity of the hole, then increases to highest and hold the maximum for a "plateau region", then attenuates gradually. But in term of cell size, for Model 1, there exist two "steep valleys"; while in Model-2, there is a "flat valley" in the immediate proximity of the hole, followed by a jump to a "plateau region" extending to the outer edge.

b) For Model-3 the relative density results show a "ladder" distribution which holds its maximum in the immediate proximity of the hole; the unit cell size increases from its minimal value in the immediate proximity of the hole, to its maximal value gradually, and maintain the maximum "plateau region" towards to outer edge. There exists a little "valley "in the middle of the "plateau region".

c) We can see from Fig. 3.5 to Fig. 3.7 that the optimization results converge with the increasing number of elements and show little mesh dependency.

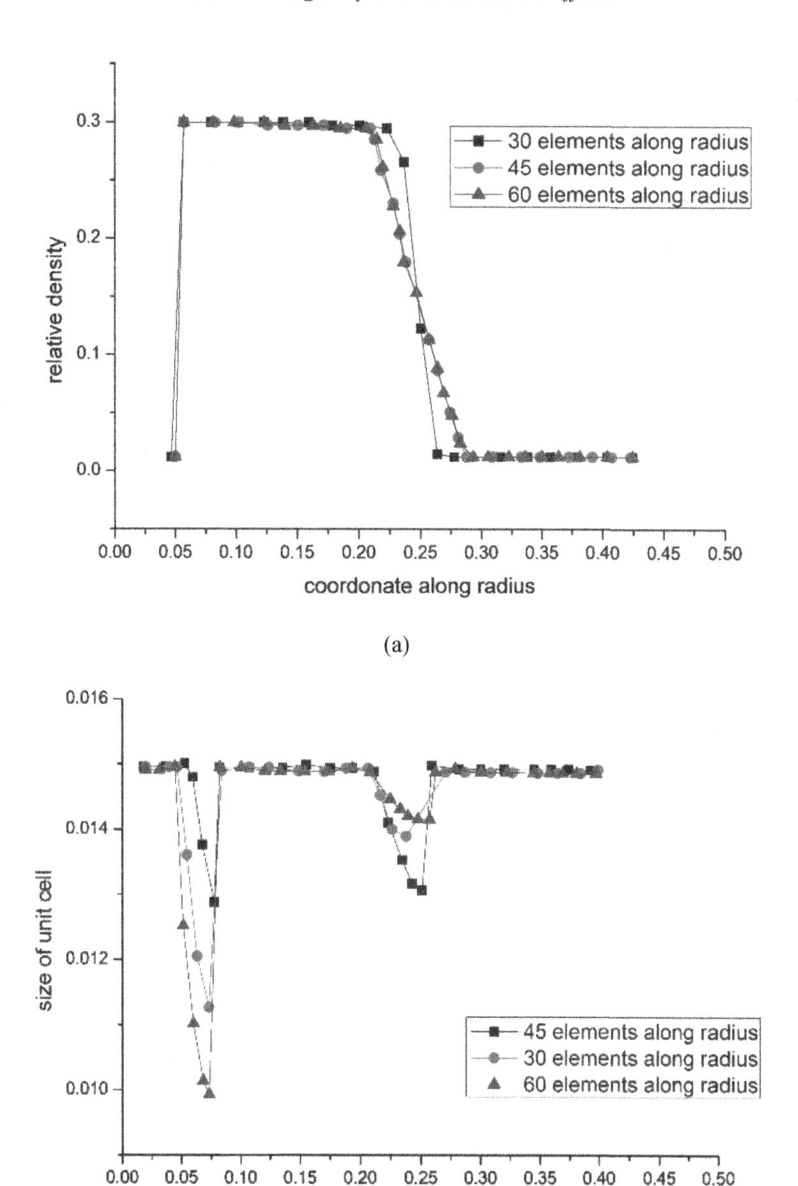

Fig. 3.5. Relative density and cell size distribution after optimization for Model-1. (a) Relative density; (b) Size of unit cell.

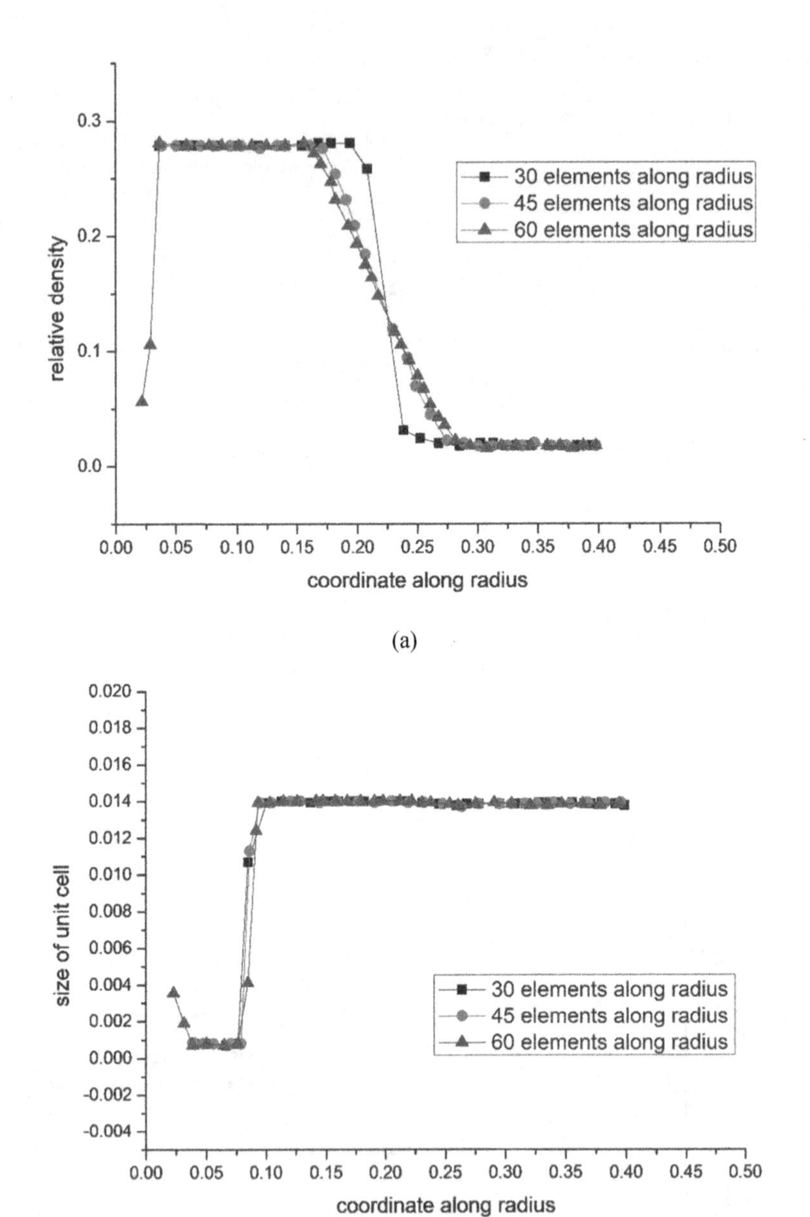

Fig. 3.6. Relative density and cell size distribution after optimization for Model-2. (a) Relative density; (b) Size of unit cell.

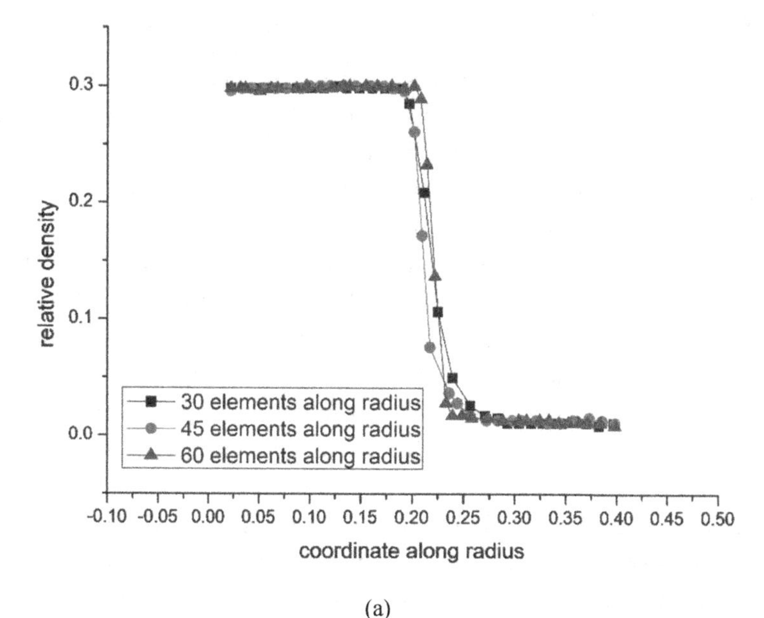

(a)

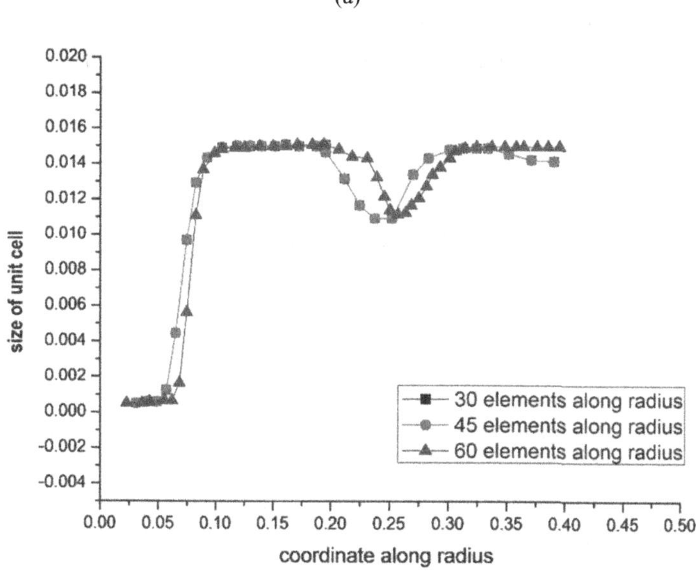

(b)

Fig. 3.7. Relative density and cell size distribution after optimization for Model-3. (a) Relative density; (b) Size of unit cell.

(2) Example 2: Influence of material ply angle on optimum results

The ply angle has an obvious influence on the stress distribution in structures. In this example, the optimum material distribution for Model-1 is investigated with material ply angle introduced as an extra design variable based on the above studies. Sensitivity of objective function to ply angle is computed with the finite difference method. And the lower and upper bounds for ply angle are 0 degree and 45 degrees respectively due to the symmetry of square cell. Other hypothesis and computation procedures are same as those above. Comparison of stress in uniform design and optimum design for cases of 45 elements along radius are given in Table 3.2 and the effects of different initial ply angle on the optimum results are also investigated. Fig. 3.8 shows the optimum distributions for three meshes composed of 30, 45 and 60 elements along radius respectively when the initial material ply angle is set to 30 degrees.

We can observe from Fig. 3.8 that the optimum material relative density distribution in this example shows the weak—strong—weak material distribution trend. However, there exist two "peaks" in the distribution of the unit cell size. And the added ply angle variable doesn't change the mesh independence of the optimization.

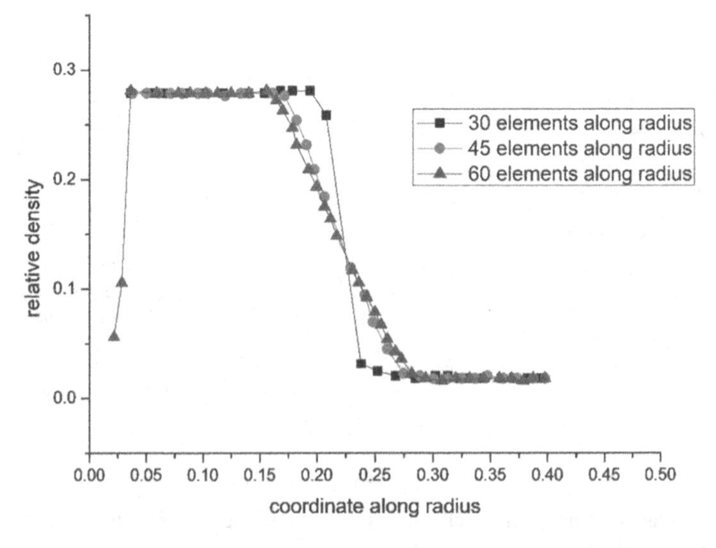

(a)

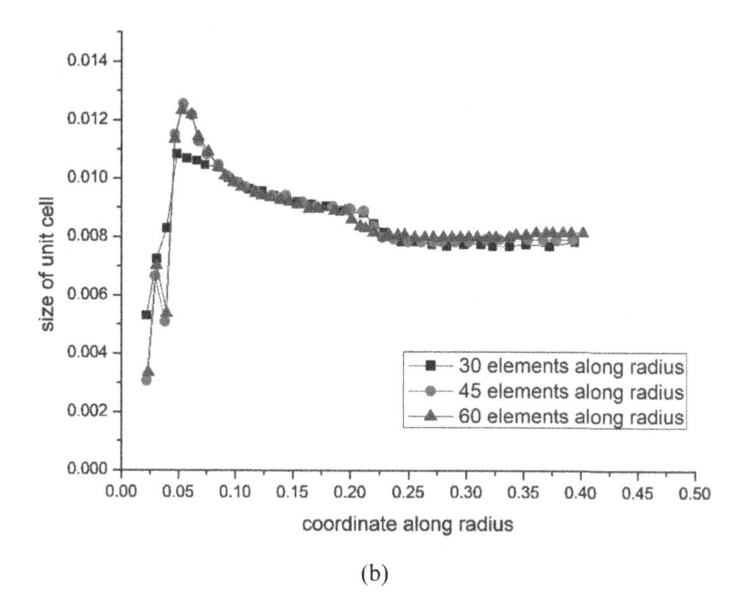

(b)

Fig. 3.8. Relative density and cell size distribution after optimization for Model-1 when the initial ply angle is set as 30 degrees

Table 3.2. Comparison of stress between uniform design and optimum design when the ply angle is introduced as a design variable for Model-1

Model-1	Ply angle (degree)	Objective function value	$\sigma_{X_2}^{P}$
Before optimization	30	6.32	2.513
After optimization	36.8	1.32×10^{-4}	0.011
Before optimization	20	57.03	7.552
After optimization	36.3	1.85×10^{-4}	0.014

(3) **Example 3:** Stress optimization around a hole in a square plate

The optimization formulation for Model-1 is carried out to optimize the stress around a hole in the square plate which is known as a classic problem with high strain gradient near the hole. The lattice plate is made of with square unit cells of local orientation y_1-y_2 coinciding with the global coordinates X_1-X_2. Due to symmetry, only a quarter of the plate shown in Fig. 3.9 needs to be analyzed. On vertical edge, displacement in the X_1-direction and the micro-rotation ϕ are constrained, whereas on horizontal edge, displacement in the X_2-directionand the micro-rotation is constrained. The edge of the hole is a traction free boundary, while the top edge has prescribed uniform pressure $F = 1$. $\boldsymbol{\sigma}_{X_2}^{P}$ is also chosen as objective stress to reduce in this example. Comparison of stress in uniform design and optimum design are given in Table 3.3 for Model-1, where the quarter plate is meshed with 1485 elements. Fig. 3.10 gives out optimization results of relative density and the size of the unit cell. (In this example, $0 < \underline{\rho} \leq \rho_i \leq 1.0$.)

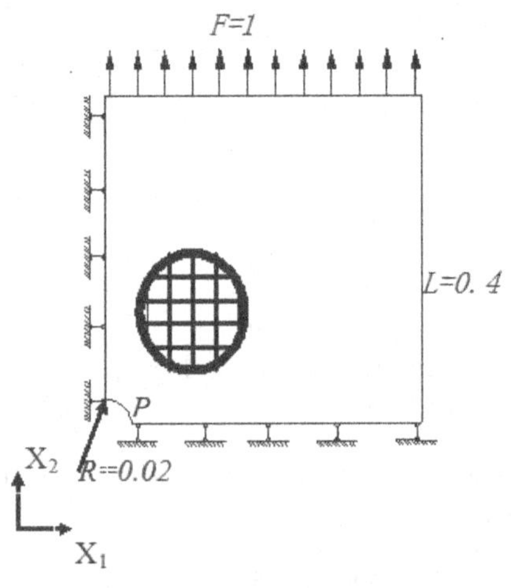

Fig. 3.9. Stress concentration around the hole in a square plate (quarter of plate is considered due to symmetry

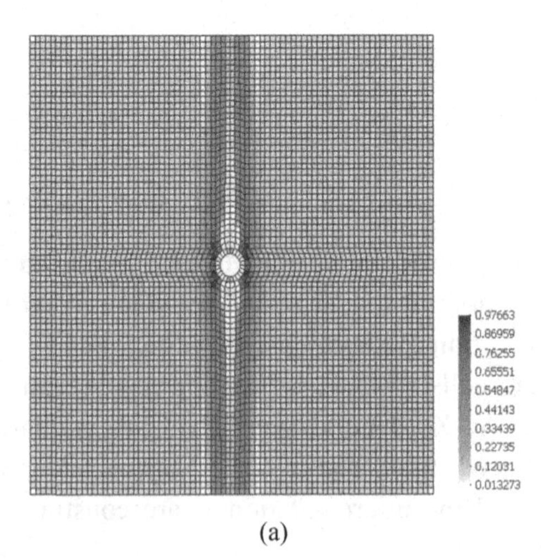

(a)

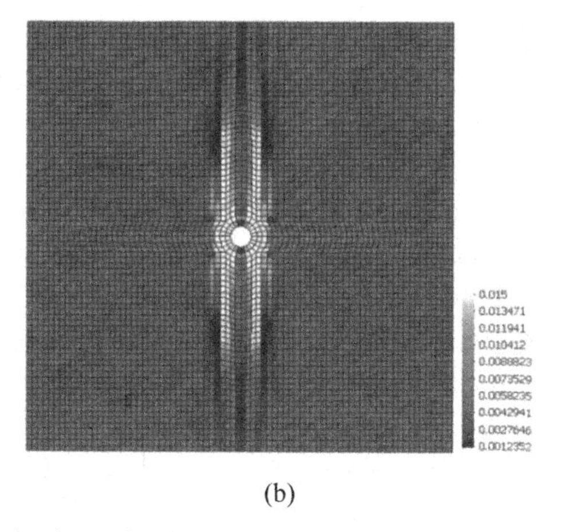

(b)

Fig. 3.10. Relative density and cell size distribution after optimization for the example of the square plate. (a) Relative density; (b) Size distribution.

Table 3.3. Comparison of stress around a hole in a square plate between uniform design and optimum design for Model-1

Model	Mesh	Initial objective function value and stress around the hole	Objective function value and stress around the hole after optimization
1	1485 elements	52.13 (stress around the hole 7.22)	6.52E-03 (stress around the hole 0.08)

As shown in Fig. 3.10, in term of relative density, it is lowest in the immediate proximity of the hole, then becomes highest along X_2-direction in the range of two to three-layer elements, then attenuates to lower value and maintain still along the edges of the square plate. This distribution can be viewed as relatively weak material being reinforced by two "strengthening ribs" right outside of the hole. In term of unit cell size, it is largest around the hole, and then decreases to minimum from the center of the hole to the top/bottom edge of the plate in the range of three-layer elements along X_2-direction. And small cell sizes distribute in most range of the plate. Due to the small effective Poisson ratio of the lattice structure with a square unit cell, the load in X_2-direction has little effect on X_1-direction. Thus, this design of "strengthening rib" is reasonable since it carries more loads in X_2-direction, hence guarantees a low stress level around the hole. This strategy (strengthening rib arranged around the hole) coincides with that adopted in engineering practice to reduce the stress concentration around a hole.

Fig. 3.11 shows the local microstructure distribution based on the optimization results (shown in Fig. 3.10) on a quarter of the plate. The cell size and cell wall thickness are not modeled exactly pointwise, but approximate results based on the optimization results. However, when we establish this approximate discrete model (each cell wall is modeled as beam elements of which properties is based on the optimization results) in MSC.NASTRAN 2005, we find that the axial stress in the beam element at the point P in Fig. 3.9 reduces from $\sigma_{X_2} = 138.8$ (obtained from the same beam when the lattice material is homogeneous distribution) to $\sigma_{X_2} = 3.86$. And at the same time, the maximum axial stress in this optimized discrete model is reduced to $\sigma_{X_2}^{Max} = 27.1$.

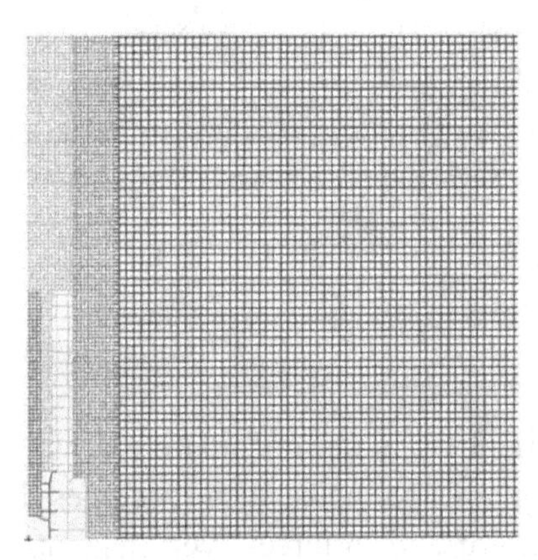

Fig. 3.11. Local micro-structures distribution based on optimization results on a quarter plate

3.2 *Lightweight design of lattice structure under the stress constraints based on EMsFEM method*

In the following, the lightweight design of lattice material based on EMsFEM method is performed under the stress constraints. However, as aforementioned in Chapter 2, one of the advantages of the EMsFEM method is the consideration of size effect in the multiscale analysis. In the practical analysis, the size effect may be significantly presented when an inappropriate boundary condition is employed or the microstructure size compared to the macrostructural size cannot be regarded as the infinitely small one. The omitting of size effect may lead to a misestimation of structural responses. Based on this consideration, the size effect is firstly investigated below.

3.2.1 *Statement of size effect in the multiscale analysis of lattice materials*

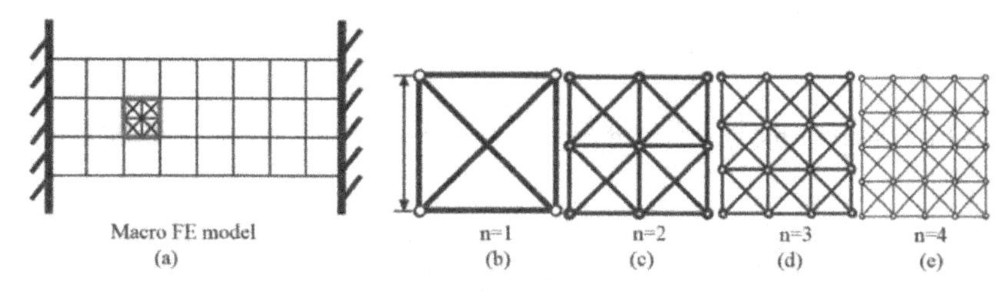

Fig. 3.12. Macro FE model and microstructure in each macro element; b–e Truss unit cells with different size factor: (b) $n = 1$; (c) $n = 2$; (d) $n = 3$; (e) $n = 4$

Four configurations of truss unit cell are considered, as shown in Fig. 3.12(b)-(e). Fig. 3.12(a) gives the finite element mesh for a macro-structure composed of lattice materials, where each macro element is uniformly built up by the truss unit cell. Fig. 3.12(b) can be considered as the basic sub-unit cell, and the truss unit cell in Fig. 3.12(c-e) is composed of a series of repetitive basic sub-unit cells. To ensure the volume of the base material in the truss unit cells being identical, the length and cross-sectional area of each rod within the truss unit cell become $1/n$ of its original values. n is named as size factor and the value 1, 2, 3 and 4 will be used in Fig. 3.12(b) to Fig. 3.12(e), respectively. If the volume of the base material is fixed, the structural response will inevitably be different as the varying of n, which is called the size effect.

Numerical examples: Corresponding discussion of the size effect

As shown in Fig. 3.13, a cantilever structure which is composed of $\mathrm{n_x} \times \mathrm{n_y} = 30 \times 6$ periodic truss unit cells, where n_x and n_y denote the number of truss unit cells in the X- and Y-directions, respectively. The left side of the cantilever is fixed, and a uniformly distributed line load $\mathrm{F} = 1000$ is applied on the right edge. Fig. 3.14(a) to Fig. 3.14(c) show three basic sub-unit cells with the same base material volume to constitute the above cantilever structure. Their cross-sectional areas have been marked on the Figures.

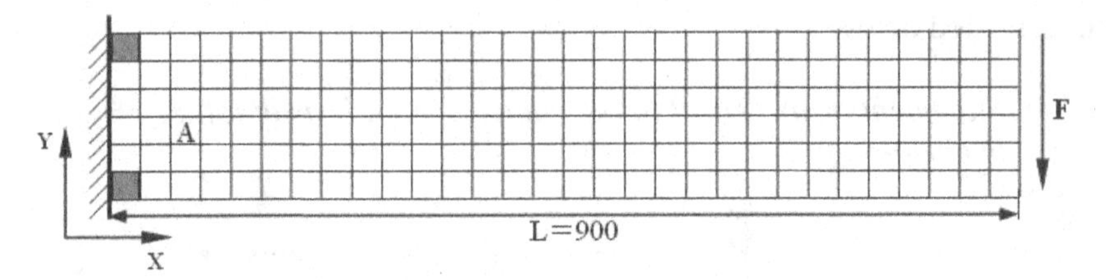

Fig. 3.13. Macroscopic FE model of the cantilever structure and its boundary conditions

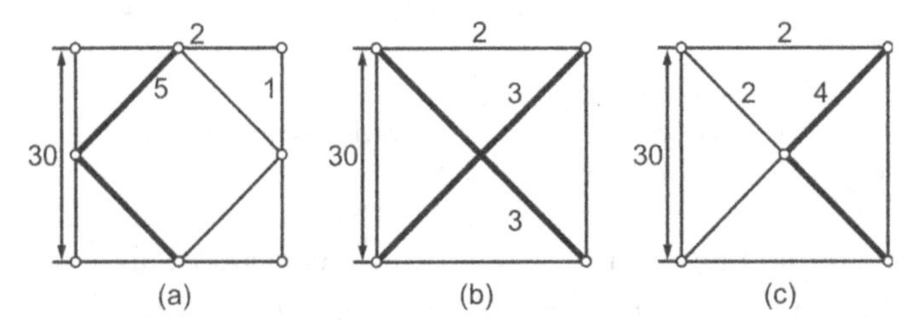

Fig. 3.14. Basic micro truss unit cells within a macro element

To study the size effect of micro unit cell composed of the above three kinds of basic sub-unit cells, the displacement of the neutral layer of the cantilever structure and the maximum axial stress of rods within the micro unit cells are analyzed with the EMsFEM by considering the increase of size factor n. And the results obtained from accurate model with finite element analysis, in which each micro component in the cantilever structure directly modeled with rod element, are provided as the reference solution. The descriptions for the following legends are given as follows. Flags 'FEM-F' and 'FEM-FA' denote the displacement results of points on the neutral layer and the maximum axial stress of the macro element A in Fig. 3.13 obtained by accurate finite element method, respectively. Similarly, the corresponding displacement and stress results obtained by EMsFEM are denoted by flags 'EMs-P' and 'EMs-PA', respectively. The corresponding displacements and stress results obtained by AH method are denoted by flags 'Homo-P' and 'Homo-A', respectively. Meanwhile, the flags ‘a’, ‘b’ and ‘c’ added behind above flags denote that the basic sub-unit cell shown in Fig. 3.14(a), (b) and (c) are used in the structural analysis, respectively.

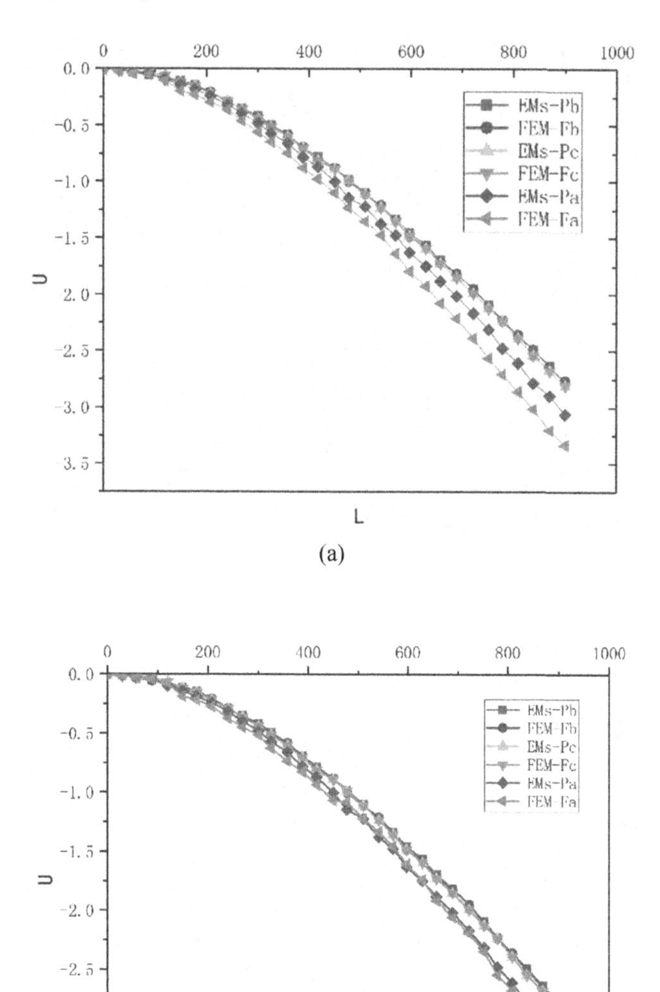

(a)

(b)

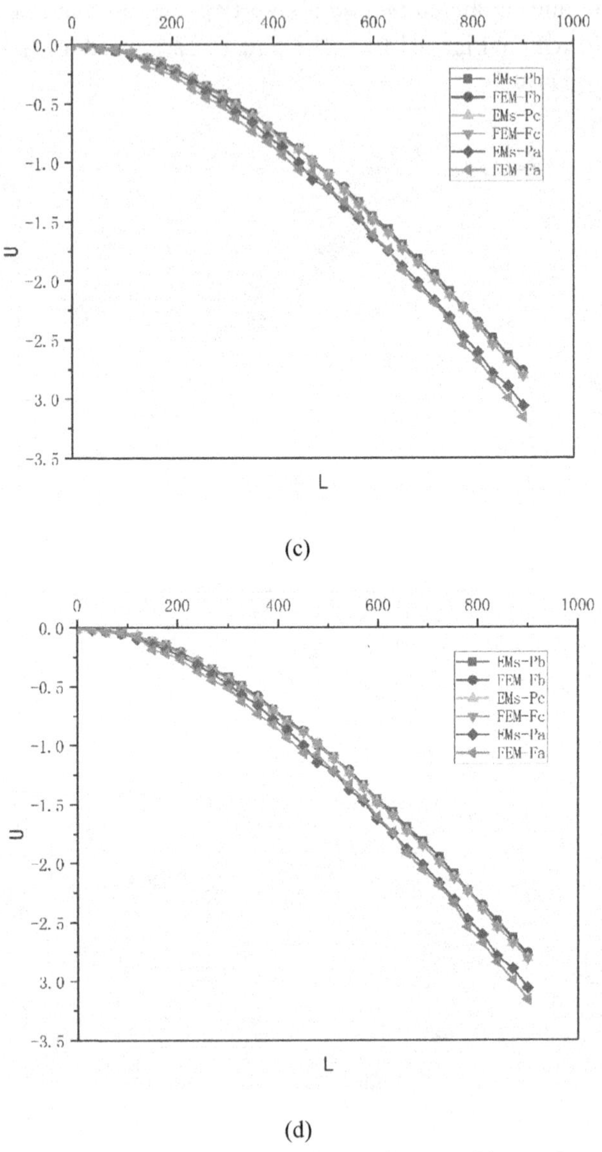

(d)

Fig. 3.15. The Y-direction displacement of points on the neutral layer of cantilever beam for different size factors. (a) $n = 1$; (b) $n = 2$; (c) $n = 3$; (d) $n = 4$

Fig. 3.15(a)–(d) show the Y-direction displacement of points on the neutral layer of cantilever beam with different size factors n. It is found that for the same volume of base material, the deformation of the cantilever composed of basic sub-unit cell shown in Fig. 3.14(b) is the minimum, and the deformation corresponding to the basic sub-unit cell shown in Fig. 3.14(a) is the maximum. The configurations of basic sub-unit cell shown in Fig. 3.14(b) and Fig. 3.14(c) are almost the same, but the material distribution of basic sub-unit cell in Fig. 3.14(b) is much more uniform. Thus, the deformation is found to be slightly smaller than that of basic sub-unit cell of Fig. 3.14(c). In the configuration of basic sub-unit cell of Fig. 3.14(a), there is a middle node on the outer edges which means that the degree of freedom on boundary increases. And an unstable quad exists in the configuration as well. So, it is found that its stiffness decreases accordingly, and the deformation is larger than the other two sub-unit cells.

From the figures with different size factors n, we can find that EMs-P and FEM-F solutions for basic sub-unit cell of Fig. 3.14(a) are obviously different when size factor $n = 1$ and the maximum error between their displacement reaches 7.7%. The reason can be attributed to the characters of the configuration of unit cell of Fig. 3.14(a), that is, there exists a middle node on the outer edges and the non-uniform distribution of the material on the inner rods. In the analysis based on EMsFEM, the boundary nodes of the unit cell cannot simultaneously meet the requirements of the balance of nodal forces and the deformation compatibility. EMs-P will show relatively large analysis error in the case. However, EMs-P and FEM-F solutions will approach each other gradually as the increase of size factor n and the maximum displacement error accounts for only 2.2% when size factor $n = 4$. The reason is that the proportion of strain energy of the boundary rods w.r.t the entire structural strain energy decreases as the increase of the number of basic sub-unit cells included in one macro element. And the boundary effect gets weaker and weaker. Thus EMs-P solution becomes more and more precise. For the basic sub-unit cell shown in Fig. 3.14(b) and Fig. 3.14(c) with only two nodes on its boundary, EMs-P and FEM-F solutions are almost the same when size factor n=1 regardless of whether the cross-sectional area of internal rods is uniform.

Fig. 3.16 shows the curve of the maximum displacement of points on the neutral layer of cantilever beam versus the size factor n. It is found that the displacement solutions obtained from EMs-P increase as the increase of the size factor and soon converge to a stationary value for the above three kinds of basic sub-unit cells. The reason is that material distributed on the upper and lower boundaries of the micro truss unit cell with high load-bearing efficiency will be decreased with the increase of size factor n, which reduces the stiffness of the cantilever structure to resist the force applied on the right side of the beam. Solutions from the EMs-P and FEM-F are essentially coincident for the basic sub-unit cells shown in Fig. 3.14(b) and Fig. 3.14(c). However, the displacements from EMs-P are smaller than those from FEM-F for basic sub-unit cell shown in Fig. 3.14(b). Because we assume periodic deformation mode in the process of constructing numerical shape functions of EMsFEM, it strengthens the stiffness of the structure numerically and leads to a relatively smaller deformation. However, the error decreases gradually with the increase of size factor n, and the maximum displacement errors only 1.4% when size factor is increased to 6. Moreover, it is also found that no size effect can be observed in the analysis using AH method, which is in line with the description of size effect in former sections.

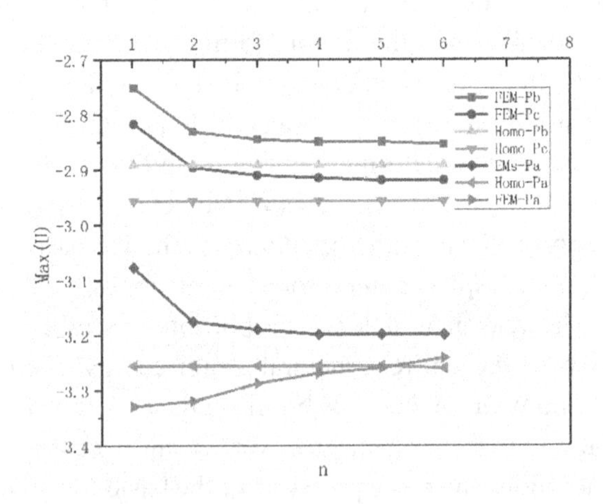

Fig. 3.16. Curve of the maximum displacement of points on the neutral layer of cantilever beam versus the size factor n

Fig. 3.17 shows the maximum axial stress of the microrods within the truss unit cell as the increase of size factor n for the above three kinds of basic sub-unit cells. The maximum axial stress corresponding to basic sub-unit cell of Fig. 3.14(a) is the largest among the three basic sub-unit cells.

And the basic sub-unit cell of Fig. 3.14(c) is the second, which is consistent with the conclusions drawing from Fig. 3.15 about displacement comparisons. For the basic sub-unit cell shown in Fig. 3.14 (a), FEM-F solutions are essentially unchanged with the increase of size factor n, and EMs-P solutions get gradually larger and converge to FEM-F solutions with the increase of size factor n. For the basic sub-unit cells shown in Fig. 3.14(b) and Fig. 3.14(c), solutions from FEM-F and EMs-P are almost the same, and their values become stationary gradually with the increase of size factor n.

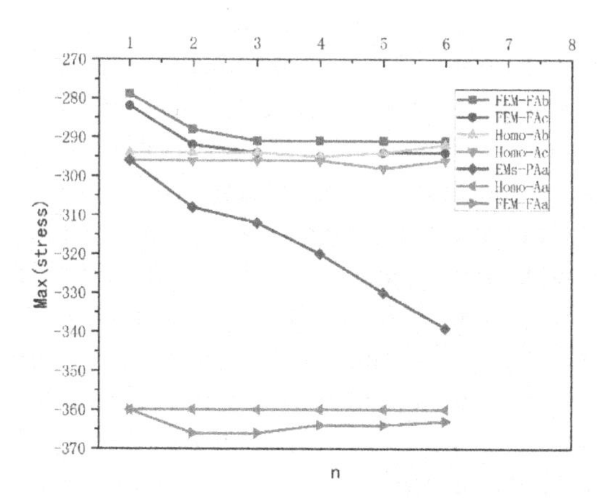

Fig. 3.17. Curve of the maximum axial stress of the macro element A versus the size factor n

Variations of the displacement of points on the neutral layer of the cantilever beam and the maximum axial stress of micro unit cells reflect the influence of the size effect of micro structures of lattice materials with the increase of size factor n. In addition, with increasing number of basic sub-unit cells in one macro element, the dimensions of basic sub-unit cells are getting smaller and smaller. The size effect of lattice material gradually decreases, and EMs-P solutions finally converge to the limit corresponding to that obtained via AH method. Based on the discussions above, we can summarize the applicable scope of EMsFEM from the above numerical experiences. FEM-F is suitable for analyzing structures composed of lattice material whose length of micro-rod is relatively large (the number of micro truss unit cells is relatively small). Generally, it can be regarded as a reference solution to validate the accuracy and efficiency of multiscale analysis methods. EMsFEM is suggested to be adopted when the length of micro-rod reduces, and the number of micro unit cells is relatively large. AH method is usually used to acquire equivalent material properties when the length of rods is very small, and the number of micro unit cells becomes huge.

3.2.2 *Minimum weight design of the structure composed of lattice materials subject to micro stress constraints*

The lightweight design of lattice structures contributes to the better implementation and improvement of its functionality [129]. Due to the lack of reliable and efficient algorithms to calculate the micro stress of lattice materials, most of the previous studies focused on the optimization design of structural stiffness i.e., compliance. Based on the research on size effect above, the minimum weight design of the structure involving complex shape and boundary conditions and composed of lattice materials is studied in this section with down scaling computing technology of EMsFEM (See 2.2.1). Again, we assume homogeneity of the microstructure of lattice materials in macroscale to meet the requirements of manufacturing cost, the cross-sectional area and axial stress of the micro-rods within the micro unit cells are introduced as the design variables and constraints, respectively, in the established optimization model. The optimization formula can be defined as follows:

$$
\begin{aligned}
&\text{find} \quad A_i, i = 1,2,\cdots \mathrm{M} \\
&\min \quad W = \mathrm{N}\sum_{i=1}^{\mathrm{M}} \rho l_i A_i \\
&\text{Constrant I} \quad -\sigma_{\mathrm{s}} \leq -\sigma_{ij}(A) \leq \sigma_{\mathrm{s}} \\
&\qquad\qquad\qquad j = 1,2,\cdots \mathrm{Q}, \mathrm{i} = 1,2,\cdots \mathrm{M} \\
&\text{Constrant II} \quad A_{\mathrm{L}} \leq A_i \leq A_{\mathrm{U}}
\end{aligned}
\tag{3.34}
$$

where A_i and l_i denote the cross-sectional area and the length of ith rod within truss unit cell, respectively. M and N are the number of rods in the truss unit cell and four-node element in macro-scale, respectively. Q is the number of macro elements selected to consider the stress constraints of the microrods. W, ρ, A_{U} and A_{L} represent weight of the structure, density of rods, the upper and lower bound of the design variables, respectively. σ_{ji} describes the axial stress of the ith rod in the jth macro element, and σ_s is the yield stress of the base material.

As shown in Fig. 3.13 cantilever structure is composed of basic sub-unit cells shown in Fig. 3.14(b) with the same boundary and load. And the influence of size effect on optimization results is discussed based on EMsFEM in this section. $\rho = 1.0$, $A_U = 3.0$, $A_L = 0.001$, and $\sigma_s = 1000$ are chosen in the example. According to the structural and loading characters, the maximum tensile and compressive axial stresses of the structure locate respectively in the upper and lower elements of the left side near the fixed boundary in the macroscale qualitative analysis. Therefore, the two elements with gray color in Fig. 3.13 are selected to calculate the axial stress of micro-rods. SQP is implemented for the optimization.

Table 3.4. Optimized results with different size factors n

n	2	3	4
Configuration	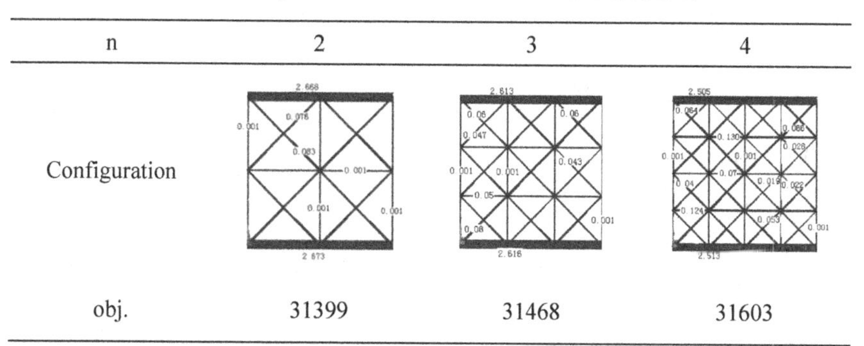		
obj.	31399	31468	31603

Table 3.4 shows the optimized results of micro truss unit cells and the optimized weight of cantilever structure by considering stress constraints with the increase of size factor. In the optimized configuration of micro truss unit cells, the cross-sectional area of the horizontal bar on the upper and lower boundaries are much larger than those of internal diagonal rods, and the cross-sectional area of vertical rods are the least and close to the given lower limit (i.e., reaching the topology optimization of the microstructure of the lattice materials), which means that they can be deleted in the actual manufacturing process. From the analysis of the mechanical characteristics of cantilever structure, it is obvious that the

horizontal normal stress plays a dominant role and the shear stress in the diagonal direction cannot be ignored too, and the vertical normal stress of cantilever structure is relatively small. Therefore, the base material is mainly distributed in the rods of upper and lower bounds to bear the corresponding normal stress, and less part of the material is allocated in internal diagonal rods to resist the shear stress. Because of less load-bearing condition of the vertical rods, the optimized cross-sectional areas of vertical rods attain the lower limit. The consistency of theoretical analysis of cantilever beam with optimization results verifies to some extent the reasonableness of the present optimization design. Table 3.4 reveals that the objective functions get slightly larger with the increase of size factor, which is consistent with the conclusions drawn from size effect. According to Fig. 3.17, the maximum axial stress of rods within the truss unit cell gradually increases with increasing number of basic sub-unit cells. Therefore, additional material is needed to meet the stress constraints with the same material strength.

3.3 *Concluding remarks*

In the chapter, two optimization problems, i.e. stress optimization based on micropolar elasticity equivalence and lightweight design with stress constraints based on EMsFEM are studied to improve the security of the lattice structures. Some important conclusions can be drawn as follows.

(1) In the stress optimization, three optimization models are proposed (*Model-1: reducing the stress along a hole boundary, Model-2: minimizing the maximum stress in specified points set, Model-3 maximizing the strength reserve at the critical area of structures*). A concurrent optimization formulation based on micropolar elasticity equivalence is proposed to carry out the optimization of the micro and macro structures. Optimum material distribution in the two scales for all three cases are obtained, and it shows that stress after optimization is lowered by a large ratio as compared to the design with uniform material distribution. Moreover, the optimized result is validated by the DNS.

(2) The size effect existing in multiscale analysis are discussed in detail, and some insight suggestions have been given for practical analysis process. The DNS FEM analysis is time-consuming and only suitable for

the analysis of a structure composed of small number of unit cells. Importantly, it is often employed as a reference solution to validate the equivalent results obtained by multiscale analysis methods. EMsFEM can well reflect the size effect. As the number of unit cells increases, the analysis results obtained by EMsFEM are gradually approach that of DNS FEM analysis. When the microstructural size is finitely small, EMsFEM is recommended to ensure an efficient and accurate analysis. Moreover, no obvious size effect is observed in the analysis of the AH method. The analysis results of DNS and EMsFEM methods are both close to that of AH method, when large number of unit cells are contained in the structure. So, AH method is suitable for cases that the size of unit cell is much smaller than the scale of macro structure. Lastly, the minimum weight design of structure composed of periodic lattice materials is discussed by considering the stress constraints of micro-rods, where is the downscaling method in EMsFEM is utillized. Optimized results show that the objective functions increase slightly with the increase of the number of basic sub-unit cells for the same material strength.

Chapter 4

Dynamics Optimization

Ultra-light cellular materials exhibit high stiffness/strength to weight ratios and bring opportunity for multifunctional performance. One of their potential applications is to build structures with optimum dynamic performance, which is extremely important for some structural parts in practical engineering and attracts a great attention. With the framework of PAMP, a concurrent multiscale optimization framework based on AH method is proposed. It aims at finding the optimal configurations of the macro-structure and material microstructure for maximizing fundamental frequency with a specific base material amount.

Microstructure of materials is assumed to be homogeneous at the macro scale to reduce the manufacturing cost. In the optimization formulation, macro and micro densities are introduced as the design variables for the macro structure and material microstructure independently. The design of microstructure is concurrently optimized with the structural topology design at the macroscale. Penalization approaches are adopted at both scales to ensure distinct topologies, i.e., SIMP at the microscale and PAMP at the macroscale. Optimizations at two scales are integrated into one system with AH theory and the distribution of base material between two scales can be determined automatically by the optimization model.

The organization of the rest part is as follows. The two-scale design model and formulations for minimizing the fundamental frequency of the structure with homogeneous cellular material is described first. It is well known that in dynamic topology optimization by the density approach such as SIMP, local vibration mode with very low vibration frequency in the low density area brings difficulty of convergence. To overcome the difficulty, special polynomial interpolation scheme for topology optimization of vibrating structures is developed based on the constraint continuity analysis. Further, the nonlinear volume preserve Heaviside filter which reduces the greyness of intermediate design and stabilizes the iteration is introduced. Formulations for the vibration of perforated plate is also interpreted in brief. Then, several numerical examples are carried out to validate the proposed method. Finally, this chapter is closed by an insight discussion.

The work of this chapter is mainly related with references [27, 130].

4.1 *Formulations of multiscale optimization for maximizing structural fundamental eigenfrequency based on PAMP*

The problem of concurrent multiscale topology optimization for maximizing the fundamental frequency of a structure can be formulated as follows:

$$\text{find} \quad \boldsymbol{X} = \{\boldsymbol{P}, \rho\} \tag{4.1}$$

$$\text{obj.} \quad \max\left\{\min_{j=1,\cdots,J}\{\lambda_j = \omega_j^2\}\right\} \tag{4.2}$$

$$\text{s.t.} \quad \begin{aligned} & \boldsymbol{K}\phi_j = \lambda_j \boldsymbol{M}\phi_j, j = 1, \dots, \mathrm{J} && \text{(a)} \\ & \phi_j^{\mathrm{T}}\boldsymbol{M}\phi_k = \delta_{jk}, k, j = 1, \dots, \mathrm{J} && \text{(b)} \\ & \varsigma = \frac{\sum_{i=1}^{\mathrm{NE}} \rho_i^{\mathrm{MA}} v_i^{\mathrm{MA}}}{V^{\mathrm{MA}}} \le \bar{\varsigma} && \text{(c)} \\ & \rho^{\mathrm{PAM}} = \frac{\sum_{i=1}^{n} \rho_l v_l^{\mathrm{MI}}}{V^{\mathrm{MI}}} = \overline{\varsigma^{\mathrm{MI}}} && \text{(d)} \\ & 0 \le \underline{P} \le P_i \le \overline{P}, \ \ i = 1, \dots, \mathrm{NE} && \text{(e)} \\ & 0 \le \underline{\rho} \le \rho_l \le \overline{\rho}, \ \ l = 1, \dots, \mathrm{n} && \text{(f)} \end{aligned} \tag{4.3}$$

In Eq. (4.2), ω_j^2 and ϕ_j denote the jth structural eigenfrequency and corresponding eigenvector, respectively. Eq. (4.3) is the governing equation of structural natural vibration. $\boldsymbol{K}$ and $\boldsymbol{M}$ are the positive definitely symmetric stiffness and mass matrices of the macro structure. Eq. (4.3) imposes the $\boldsymbol{M}$ orthonormalization condition on eigenvectors.

δ_{jk} is Kronecker's delta. In the optimization of maximum natural frequency by topology optimization, the order of the modes could change during optimization, namely so called mode switching [131-133], thus J candidate frequencies are considered in the objective function Eq. (4.2). The eigenvalue problem is solved using the subspace iteration method.

The constraint Eq. (4.3) (c) sets the upper bound of the available base material, $\bar{\varsigma} \cdot V^{\mathrm{MA}}$ is prescribed base material amount, $\bar{\varsigma}$ represents the fraction of prescribed base material on the total base material for filling up the whole macro design domain using solid material. Since an element in the macroscale is composed of cellular material unit cells, $\rho_i^{\mathrm{MA}} = P_i \times \rho^{\mathrm{PAM}}$ is the ith element volumetric density, where ρ^{PAM} is the relative density of cellular material unit cell and its definition is given in Eq. (4.3) (d). v_l^{MI} denotes the volume of l-th micro element. V^{MI} is the total volume of cellular material unit cell. $\overline{\varsigma^{\mathrm{MI}}}$ is the prescribed relative density of the cellular material. Eq. (4.3) (e) and Eq. (4.3) (f) give the lower and upper limits of macro and micro design variables, respectively. To avoid singularity in computation, low limit 0.001 is specified for both macro and micro volumetric material densities $\boldsymbol{P}$ and ρ.

The global stiffness matrix $\boldsymbol{K}$ and mass matrix $\boldsymbol{M}$ can be calculated by

$$\boldsymbol{K} = \sum_{i=1}^{\mathrm{NE}} \int_{\Omega^{\mathrm{e}}} \boldsymbol{B}^{\mathrm{T}} \times \boldsymbol{D}^{\mathrm{MA}} \times \boldsymbol{B} d\Omega^{\mathrm{e}} = \sum_{i=1}^{\mathrm{NE}} \boldsymbol{K}_i^{\mathrm{MA}} \tag{4.4}$$

$$\boldsymbol{M} = \sum_{i=1}^{\mathrm{NE}} \int_{\Omega^{\mathrm{e}}} \rho^{\mathrm{MA}} \times \eta \times \boldsymbol{N}^{\mathrm{T}} \boldsymbol{N} d\Omega^{\mathrm{e}} = \sum_{i=1}^{\mathrm{NE}} \boldsymbol{M}_i^{\mathrm{MA}} \tag{4.5}$$

where $\boldsymbol{K}_i^{\mathrm{MA}}$ and $\boldsymbol{M}_i^{\mathrm{MA}}$ represent respectively the ith element stiffness and mass matrices in the form of expanded structural global degree of freedom. $\boldsymbol{D}^{\mathrm{MA}}$ and ρ^{MA} are the equivalent constitutive matrix and structural density. ***B*** and ***N*** are the strain-displacement matrix and the shape function matrix at the macro scale, respectively.

The two-scale optimization problem is solved by using the SLP (Sequential Linear Programming) [134] in this chapter. Explicit expression of sensitivity is important to enhance the efficiency of the gradient-based mathematical programming algorithms such as SLP.

For the examples in this section, we trace the first three eigenfrequencies during the optimization process, and no eigenmode switching and coincidence eigenvalue phenomenons are encountered. The curves of the evolution of eigenvalues for some examples (Example 1 and Example 6) are referred to Fig. 4.9 and Fig. 4.14. Thus for all test cases presented here, the optimized fundamental eigenfrequency is unimodal and differentiable. When the *k*th eigenfrequency is unimodal, detailed sensitivity analysis of the two-scale optimization problem is given in Appendix C. However, multiple eigenfrequencies problem is very important in the optimization of vibrating structure and isn't considered in the present study. In the case of multiple eigenfrequencies, the eigenfrequencies are non-differentiable. Sensitivity analysis of multiple eigenfrequencies has been investigated extensively in many studies [131, 132, 135-137]. In order to achieve distinct topologies at both scales, penalization methods are applied. At the microscale, it is natural to utilize SIMP, a method commonly used in traditional structural topology optimization. Assuming the modulus matrix of the base material is $\boldsymbol{D}^{\mathrm{B}}$, the modulus matrix $\boldsymbol{D}^{\mathrm{MI}}$ at a point with density interpolation at the microscale can be expressed as

$$\boldsymbol{D}^{\mathrm{MI}} = \rho^{\mu} \boldsymbol{D}^{\mathrm{B}} \tag{4.6}$$

At the macroscale, since the cellular material can be anisotropic, given any porous anisotropic material with modulus matrix $\boldsymbol{D}^{\mathrm{H}}$, a point with density P has the modulus matrix $\boldsymbol{D}^{\mathrm{MA}}$ as expressed by

$$\boldsymbol{D}^{\mathrm{MA}} = P^{\alpha} \boldsymbol{D}^{\mathrm{H}} \tag{4.7}$$

By assuming the power indexes $\mu > 1$, $\alpha > 1$, the material densities are penalized for closing to either 0 or 1. In Eq. (4.7), the interpolation is based on the PAMP method, which has been proven to be well suited for the optimization for a structure made of micro anisotropic porous materials .

In terms of artificial relative density value P, the finite element mass matrix may be expressed as

$$\boldsymbol{M}_{\mathrm{e}} = P^{\xi} \boldsymbol{M}_{\mathrm{e}}^{*} \tag{4.8}$$

where $\boldsymbol{M}_\mathrm{e}^*$ represents the element mass matrix corresponding to macro element with relative density $P = 1$. $\xi = 1$ is chosen in this chapter.

4.1.1 *Polynomial interpolation scheme to eliminate the local vibration mode*

One of the main problems in the topology optimization with respect to eigenfrequencies or buckling loads using the density approach each on SIMP is the possibility of localized modes with very low values of corresponding eigenfrequencies [138, 139]. The localized modes may occur in areas with low values of the element relative densities for typical penalization values $\alpha = 3$ and $\xi = 1$. To eliminate these localized eigenmodes, many methods are presented in the field of dynamic optimization [138, 140]. Based on the constraint continuity analysis approach, Cheng and Wang [140] pointed out that the limiting value of the ratio between stiffness and mass has a great effect on the lowest eigenfrequencies when design variable P approaches to zero.

Following the conception of PAMP or SIMP and an isotropic material is considered for simplicity in the following, a generalized interpolation function $g(P)$ can be introduced to describe elastic modulus E^{MA} at a macro point with density P, i.e.,

$$E^{\mathrm{MA}} = g(P)E^{\mathrm{H}} \tag{4.9}$$

Then, the expression of $g(P)$ needs to meet the following principals.

(1) **Principal 1:** The interpolated E^{MA} should correctly reflect the absence or presence of the material when the macro element density P is taken as 0 or 1, respectively. i.e.,

$$g(P)|_{P=0} = 0, g(P)|_{P=1} = 1 \tag{4.10}$$

(2) **Principal 2**: $g(P)$ should give an effective penalization for intermedia element density to ensure a distinct 0/1 design. For the two arbitrary element densities P_1 and P_2, they satisfy the relations $0 < P_1 < P_2 < 1$. It assumes a quantity P exists in the interval (P_1, P_2), i.e., $P_1 < P < P_2$. To enable a penalization of the intermedia element density, for any $P = \alpha P_1 + (1 - \alpha)P_2$ and $0 < \alpha < 1$, the interpolation function $g(P)$ follows

$$\frac{g(P)}{P} < \frac{\alpha g(P_1)}{P_1} + \frac{(1-\alpha)g(P_2)}{P_2} \tag{4.11}$$

Obviously, Eq. (4.11) implies that $g(P)/P$ is a convex function, i.e.,

$$\frac{d^2(g(P)/P)}{dP^2} > 0 \tag{4.12}$$

(3) **Principal 3:** To illustrate the nature of structural local vibration in low element density region, the macrostructural stiffness $\boldsymbol{K}$ and mass matrix $\boldsymbol{M}$ can be rearranged as

$$\begin{aligned} \boldsymbol{K} &= \boldsymbol{K}_{\mathrm{f}} + \boldsymbol{K}_{\mathrm{d}} = \boldsymbol{K}_{\mathrm{f}} + g(P)\boldsymbol{K}_{\mathrm{d0}} \\ \boldsymbol{M} &= \boldsymbol{M}_{\mathrm{f}} + \boldsymbol{M}_{\mathrm{d}} = \boldsymbol{M}_{\mathrm{f}} + P\boldsymbol{M}_{\mathrm{d0}} \end{aligned} \tag{4.13}$$

where the subscript 'd' refers to the low density region (grey area in Fig. 4.1) and the subscript 'f' refers to the rest of the design domain, $\boldsymbol{K}_{\mathrm{d0}}$ and $\boldsymbol{M}_{\mathrm{d0}}$ are the non-penalized stiffness and mass matrix in the low density region. Based on Rayleigh Principle, the structural frequency can be estimated by

$$\omega^2 = \frac{\boldsymbol{u}^{\mathrm{T}}\boldsymbol{K}\boldsymbol{u}}{\boldsymbol{u}^{\mathrm{T}}\boldsymbol{M}\boldsymbol{u}} = \frac{\boldsymbol{u}^{\mathrm{T}}\boldsymbol{K}_{\mathrm{f}}\boldsymbol{u} + g(P)\boldsymbol{u}^{\mathrm{T}}\boldsymbol{K}_{\mathrm{d0}}\boldsymbol{u}}{\boldsymbol{u}^{\mathrm{T}}\boldsymbol{M}_{\mathrm{f}}\boldsymbol{u} + P\boldsymbol{u}^{\mathrm{T}}\boldsymbol{M}_{\mathrm{d0}}\boldsymbol{u}} \tag{4.14}$$

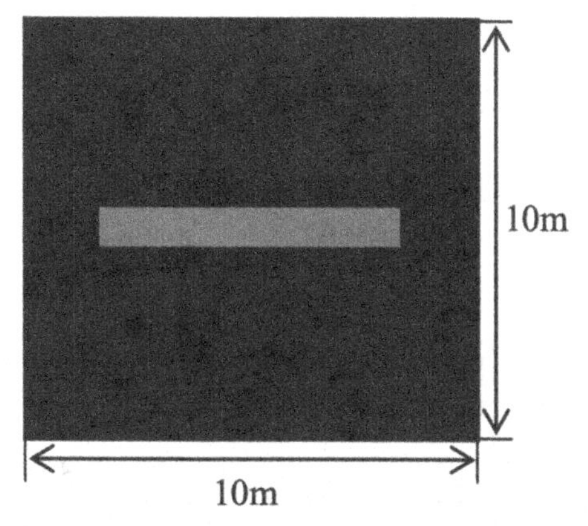

Fig. 4.1. Design (Grey) and non-design domain of a rectangular plate

The terms $g(P)\boldsymbol{u}^{\mathrm{T}}\boldsymbol{K}_{\mathrm{d}0}\boldsymbol{u}$ and $P\boldsymbol{u}^{\mathrm{T}}\boldsymbol{M}_{\mathrm{d}0}\boldsymbol{u}$ will approach zero as the element density P decrease to zero gradually.

Now, it assumes that the global vibration mode switches to a local mode occurred in the low element density region when the element density trends to zero. It means that the vibration energy of the local region become dominated over the structure. Then the Rayleigh quotient can be approximated by

$$\omega^2 = \frac{\boldsymbol{u}^{\mathrm{T}}\boldsymbol{K}\boldsymbol{u}}{\boldsymbol{u}^{\mathrm{T}}\boldsymbol{M}\boldsymbol{u}} = \frac{g(P)\boldsymbol{u}^{\mathrm{T}}\boldsymbol{K}_{\mathrm{d}0}\boldsymbol{u}}{P\boldsymbol{u}^{\mathrm{T}}\boldsymbol{M}_{\mathrm{d}0}\boldsymbol{u}} \tag{4.15}$$

It is very interesting to find that the quotient basically depends on $g(P)/P$. If $P \to 0$ such that $g(P)/P \to \infty$, then the local vibration will never be emerged. If $P \to 0$ such that $g(P)/P \to 0$, the first eigenmode must be a local one. For other cases, i.e. $P \to 0$ such that $g(P)/P$ is neither infinite nor zero, more elaborate analysis is needed. However, when $g(P)/P \to \bar{\mathrm{C}}$ and $\bar{\mathrm{C}}$ is a very small constant, then first vibration mode might be a local one. Thus, a careful selection of the value of C is important. Then, to eliminate the local vibration mode, the following relation need to be satisfied

$$\lim_{P\to 0} g(P)/P = \infty \text{ or } \bar{\mathrm{C}} \tag{4.16}$$

According to the L'Hopital's rule, the expression above can be rewritten as

$$\lim_{P\to 0}\frac{g(P)}{P} = \lim_{P\to 0}\frac{dg(P)}{dP} \tag{4.17}$$

To summary, the third principal is given below

$$\lim_{P\to 0}\frac{dg(P)}{dP} \geq \bar{\mathrm{C}} \tag{4.18}$$

(4) **Principal 4**: Moreover, the interpolated modulus is required to be located in the Hashin-Shtrikman interval, i.e.,

$$\frac{(2+P)E_1 + (1-P)E_2}{2(1-P)E_1 + (1+2P)E_2}E_2 \leq E(P) \leq \frac{PE_1 + (3-P)E_2}{(3-2P)E_1 + 2PE_2}E_1 \tag{4.19}$$

where E_1 and E_2 are the elastic modulus of the two base materials and $E_1 > E_2$. For the present study, $E_1 = E^{\mathrm{H}}$ and $E_2 = 0$. So, the Hashin-Shtrikman interval can be updated as

$$0 \leq E(P) \leq \frac{P}{(3-2P)} E_1 \tag{4.20}$$

For the lower limit of Hashin-Shtrikman interval, it requires

$$\lim_{P\to 0} \left.\frac{dg(P)}{dP}\right|_{P\to 0} \geq 0 \text{ and } \lim_{P\to 0} \left.\frac{dg(P)}{dP}\right|_{P\to 1} \leq \infty \tag{4.21}$$

For the upper limit of Hashin-Shtrikman interval, it requires

$$\lim_{P\to 0} \left.\frac{dg(P)}{dP}\right|_{P\to 0} \leq \left.\frac{d\left(\frac{P}{3-2P}\right)}{dP}\right|_{P\to 0} = \frac{1}{3} \tag{4.22}$$

$$\lim_{P\to 0} \left.\frac{dg(P)}{dP}\right|_{P\to 1} \geq \left.\frac{d\left(\frac{P}{3-2P}\right)}{dP}\right|_{P\to 0} = 3 \tag{4.23}$$

Combination of the relations with respect to the lower and upper limit of Hashin-Shtrikman interval, it finally results in

$$\begin{aligned} 0 \leq \lim_{P\to 0} \left.\frac{dg(P)}{dP}\right|_{P\to 0} \leq \frac{1}{3} \\ 3 \leq \lim_{P\to 0} \left.\frac{dg(P)}{dP}\right|_{P\to 1} \leq \infty \end{aligned} \tag{4.24}$$

In the present study, the 3-order polynomial interpolation function $g(P) = \mathrm{a}P^3 + \mathrm{b}P^2 + \mathrm{c}P + \mathrm{d}$ is constructed. According to the four principals above, it yields

$$\begin{aligned} g(P)|_{P=0} = \mathrm{d} = 0 \\ g(P)|_{P=1} = \mathrm{a} + \mathrm{b} + \mathrm{c} = 1 \end{aligned} \tag{4.25}$$

$$\frac{d^2(g(P)/P)}{dP^2} = 2\mathrm{a} > 0$$

$$\left.\frac{dg(P)}{dP}\right|_{P\to 0} = \mathrm{c} = \mathrm{C}_1 \in [\bar{\mathrm{C}}, \frac{1}{3}]$$

$$\left.\frac{dg(P)}{dP}\right|_{P\to 1} = 3\mathrm{a} + 2\mathrm{b} + \mathrm{c} = \mathrm{C}_2 \in [3, \infty]$$

Different values of C_1 and C_2 will result in different interpolation function. Here the three typical expressions are given below.

$$\mathrm{C}_1 = \frac{1}{3}, \mathrm{C}_2 = 3: \quad g_1(P) = \frac{4}{3}P^3 - \frac{2}{3}P^2 + \frac{1}{3}P$$

$$\mathrm{C}_1 = \frac{1}{10}, \mathrm{C}_2 = 3: \quad g_1(P) = 1.1P^3 - 0.2P^2 + 0.1P \tag{4.26}$$

$$\mathrm{C}_1 = \frac{1}{100}, \mathrm{C}_2 = 3: \quad g_1(P) = 1.01P^3 - 0.02P^2 + 0.01P$$

Based on the conception of PAMP, the above-mentioned polynomial interpolation can be easily extended to the anisotropic cellular material. To avoid the local vibration mode, the penalization for the element stiffness matrix is expressed as

$$\boldsymbol{D}^{\mathrm{MA}} = f(P)\boldsymbol{D}^{\mathrm{H}} \tag{4.27}$$

In the following, the second interpolation scheme shown in Eq. (4.26) will be employed in the following studies, i.e.

$$f(P) = 1.1P^3 - 0.2P^2 + 0.1P \tag{4.28}$$

Fig. 4.2 shows comparison of our polynomial penalization and the traditional exponent penalization.

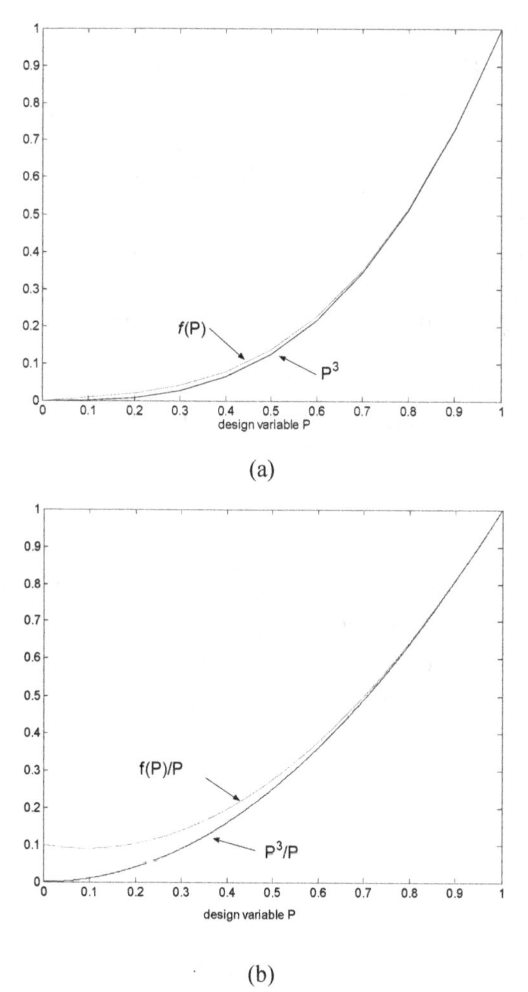

Fig. 4.2. Comparisons of polynomial penalization Eq. (4.28) and exponent penalization

With the equivalent constitutive matrix of cellular material and penalization method, the global stiffness matrix $\boldsymbol{K}$ and mass matrix $\boldsymbol{M}$ can now be calculated by

$$\begin{aligned} \boldsymbol{K} &= \sum_{i=1}^{\mathrm{NE}} \int_{\Omega^{\mathrm{e}}} \boldsymbol{B}^{\mathrm{T}} \cdot \boldsymbol{D}^{\mathrm{MA}} \cdot \boldsymbol{B} d\Omega^{\mathrm{e}} \\ &= \sum_{i=1}^{\mathrm{NE}} f(P_i) \int_{\Omega^{\mathrm{e}}} \boldsymbol{B}^{\mathrm{T}} \cdot \boldsymbol{D}^{\mathrm{H}} \cdot \boldsymbol{B} d\Omega^{\mathrm{e}} = \sum_{i=1}^{\mathrm{NE}} f(\mathrm{P}_i) \boldsymbol{K}_i^* \end{aligned} \tag{4.29}$$

$$\boldsymbol{M} = \sum_{i=1}^{\mathrm{NE}} \left(P_i \int_{\Omega^e} \eta \cdot \rho^{\mathrm{PAM}} \boldsymbol{N}^{\mathrm{T}} \boldsymbol{N} d\Omega^{\mathrm{e}} \right) = \sum_{i=1}^{\mathrm{NE}} P_i \cdot \boldsymbol{M}_i^* \tag{4.30}$$

In these equations, $\boldsymbol{K}_i^*$ and $\boldsymbol{M}_i^*$ represent respectively the *i*th element stiffness and mass matrices with macro element volumetric relative density $P = 1$.

4.1.2 *Volume preserved Heaviside projection*

To eliminate the checkerboard pattern and mesh-dependence phenomenon, linear density filtering method [141] is common utilized. The linear formation of the filter on the two scales can be written as

$$\bar{P} = \frac{\sum_{i \in \boldsymbol{S}_{\mathrm{e}}^{\mathrm{MA}}} \omega^{\mathrm{MA}}(\boldsymbol{x}_i) P_i}{\sum_{i \in \boldsymbol{S}_{\mathrm{e}}^{\mathrm{MA}}} \omega^{\mathrm{MA}}(\boldsymbol{x}_i)} \tag{4.31}$$

$$\bar{\rho} = \frac{\sum_{j \in \boldsymbol{S}_{\mathrm{e}}^{\mathrm{MI}}} \omega^{\mathrm{MI}}(y_j) \rho_j}{\sum_{j \in \boldsymbol{S}_{\mathrm{e}}^{\mathrm{MI}}} \omega^{\mathrm{MI}}(\boldsymbol{y}_j)} \tag{4.32}$$

$\overline{P}$ and $\overline{\rho}$ are the filtered design variables using linear density filter. The weighting functions $\omega^{\mathrm{MA}}(\boldsymbol{x}_i)$ [27] and $\omega^{\mathrm{MI}}(\boldsymbol{y}_i)$ are defined by

$$\omega^{\mathrm{MA}}(\boldsymbol{x}_i) = \begin{cases} \dfrac{(R - \|\boldsymbol{x}_i - \boldsymbol{x}_{\mathrm{e}}\|)}{R}, & \text{if } \boldsymbol{x}_i \epsilon \boldsymbol{S}_{\mathrm{e}}^{\mathrm{MA}} \\ 0, & \text{otherwise} \end{cases} \tag{4.33}$$

$$\omega^{\mathrm{MI}}(\boldsymbol{y}_i) = \begin{cases} \dfrac{\left(r - \|\boldsymbol{y}_j - \boldsymbol{y}_{\mathrm{e}}\|\right)}{r}, & \text{if } \boldsymbol{y}_i \epsilon \boldsymbol{S}_{\mathrm{e}}^{\mathrm{MI}} \\ 0, & \text{otherwise} \end{cases} \tag{4.34}$$

R is the given filter radius in the macro design domain, and r is the given filter radius in the micro design domain. The primary role of the filter radius is to identify the elements that influence the relative density of element e. For example, in the macro design domain we draw a circle of radius R centered at the center of element e, thus generate the circular sub-domain $\boldsymbol{S}_{\mathrm{e}}^{\mathrm{MA}}$. Elements with centers located inside $\boldsymbol{S}_{\mathrm{e}}^{\mathrm{MA}}$ contribute to the computation of relative density of element e in the macro domain. Similarly, the sub-domains $\boldsymbol{S}_{\mathrm{e}}^{\mathrm{MI}}$ is also specified by the elements that have centers within the given filter radius r of the center of the element e in the micro design domain. $\boldsymbol{x}_i$ and $\boldsymbol{y}_j$ denote the spatial (center) locations of the element i in the macro design domain and element j in the micro design domain, respectively.

Whereas it will exacerbate the elements with intermedia densities in the optimized results. Although a polynomial penalization for intermedia element density is introduced in Eq. (4.28), but numerical examples have shown that this penalization is not enough to enable a 0/1 distinct design for the micro and macro designs. Thus, some other remedies should be considered. Heaviside step function has been turned out be an effective method to enhance the distinct design. The penalized element density $\tilde{\rho}_{\mathrm{e}}$ by Heaviside step function can be expressed as a smooth function with respect to the filtered element density $\bar{\rho}_{\mathrm{e}}$ by introducing one parameter β (cf. Guest et al. [102])

$$\tilde{\rho} = 1 - e^{-\beta\bar{\rho}} + \bar{\rho}e^{-\beta} \tag{4.35}$$

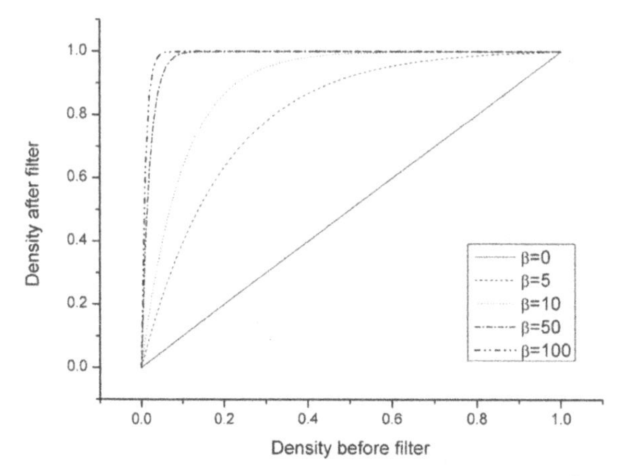

Fig. 4.3. Smoothed Heaviside function with different β

Notably, here illustrations of the Heaviside filter only for the micro density are given, and the filter for macro element density P can be performed using the similar manner. A figure illustrating Eq. (4.35) with different β is shown in Fig. 4.3. As we can see, for β equal to zero, $\tilde{\rho}$ is equal to $\bar{\rho}$, i.e. Eq. (4.35) does nothing to linearly filtered density $\bar{\rho}$. For β approaching infinity, Eq. (4.35) acts as the step function, i.e. only if $\bar{\rho}$ equal to zero will $\tilde{\rho}$ be zero, and when $\bar{\rho} > 0$, $\tilde{\rho}$ is equal to one. For other cases of β, Eq. (4.35) acts as a penalty forcing the intermediate density $\bar{\rho}$ moving to one. In practical optimization process, β is set to a small value (e.g., 0.1) at first, and increased gradually during iteration.

Sigmund [104] presented another Heaviside filter contrary to that shown in Eq. (4.35) and its smooth form is given as

$$\tilde{\rho} = e^{-\beta(1-\bar{\rho})} - (1-\bar{\rho})e^{-\beta} \tag{4.36}$$

A figure illustrating Eq. (4.36) with different β is shown in Fig. 4.4. As we can see, the modified Heaviside filter in Eq. (4.36) acts exactly contrary to the Heaviside filter in Eq. (4.35). For β equal to zero, $\tilde{\rho}$ is equal to $\bar{\rho}$, i.e. Eq. (4.36) does nothing to linearly filtered density $\bar{\rho}$. For β approaching infinity, Eq. (4.36) acts as the step function, i.e. only if $\bar{\rho}_e$ equal to one will $\tilde{\rho}$ be one, and when $\bar{\rho} < 1$, $\tilde{\rho}$ is equal to zero. For other cases of β, Eq. (4.36) acts as a penalty forcing the intermediate density $\bar{\rho}$ moving to zero. In practical optimization process, β is set to a small value (e.g., 0.1) at first, and increased gradually during iteration.

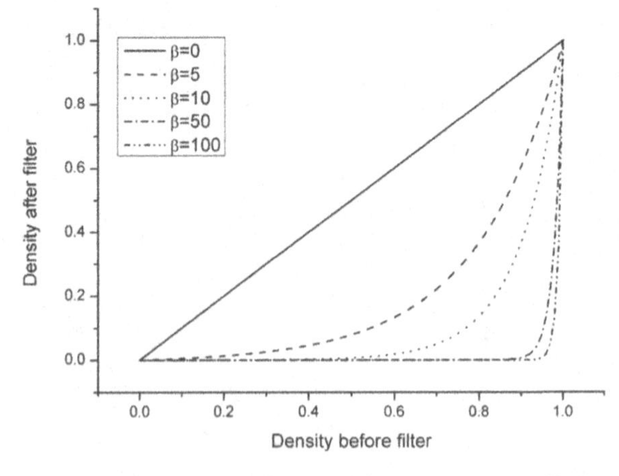

Fig. 4.4. Smoothed Modified Heaviside function with different β

Although the Heaviside smooth functions shown in Eqs. (4.35) and (4.36) is helpful to obtain a distinct topology, but some new numerical issues may appear. Obviously, it can be found that the material before and after penalization is not identical in volume. As the parameter β increases, the extra iterations are need to re-satisfy the material volume constraint. This makes iteraition convergence different. Based on this consideration, a new volume preserving Heaviside function is proposed by Xu et al. [130].

To satisfy the requirement of volume preservation, our basic idea is that the new Heaviside function should have such a property: for $0 \leq \bar{\rho} < \eta$, $\tilde{\rho}_{\mathrm{e}}$ is equal to zero; for $\eta < \bar{\rho} \leq 1$, $\tilde{\rho}$ is equal to one; and for $\bar{\rho}$ equal to η, $\tilde{\rho}$ is η. Here η is a new parameter introduced in our new Heaviside function and $\eta \in [0,1]$. The new Heaviside step function can be written as:

$$\tilde{\rho} = \begin{cases} 0 & 0 \leq \bar{\rho} < \eta \\ \eta & \bar{\rho} = \eta \\ 1 & \eta < \bar{\rho} \leq 1 \end{cases} \tag{4.37}$$

The smooth form of Eq. (4.37) can be written as

$$\tilde{\rho} = \begin{cases} \eta\left[e^{-\beta(1-\frac{\bar{\rho}}{\eta})} - \left(1-\frac{\bar{\rho}}{\eta}\right)e^{-\beta}\right] & 0 \leq \bar{\rho} < \eta \\ \eta & \bar{\rho} = \eta \\ (1-\eta)\left[1 - e^{-\beta\left(\frac{\bar{\rho}-\eta}{1-\eta}\right)} + \frac{(\bar{\rho}-\eta)e^{-\beta}}{1-\eta}\right] + \eta & \eta < \bar{\rho} \leq 1 \end{cases} \tag{4.38}$$

where there are two parameters β and η. Note that the first and third expressions of Eq. (4.38) are both equal to η for $\bar{\rho} = \eta$, which means this new Heaviside function is a continuous function. Thus, it yields

$$\tilde{\rho} = \begin{cases} \eta\left[e^{-\beta(1-\frac{\bar{\rho}}{\eta})} - \left(1-\frac{\bar{\rho}}{\eta}\right)e^{-\beta}\right] & 0 \leq \bar{\rho} \leq \eta \\ (1-\eta)\left[1 - e^{-\beta\left(\frac{\bar{\rho}-\eta}{1-\eta}\right)} + \frac{(\bar{\rho}-\eta)e^{-\beta}}{1-\eta}\right] + \eta & \eta < \bar{\rho} \leq 1 \end{cases} \tag{4.39}$$

Notably, we can further prove that the derivative $d\tilde{\rho}/d\bar{\rho}$ is continuous. Figures illustrating Eq. (4.39) with different β and η are shown in Fig. 4.5 and Fig. 4.6.

As we can see, the new Heaviside filter in Eq. (4.39) is a combination of the original Heaviside filters shown in Eqs. (4.35) and (4.36) by simply rescaling them from $[0, 1]$ to $[\eta, 1]$ and $[0, \eta]$ respectively. When η approaches 0, Eq. (4.39) is degenerated to the original Heaviside filter in Eq. (4.35), and when η approaches 1, it will be degenerated to the Heaviside filter in Eq. (4.36). For η between 0 and 1, Eq. (4.39) is a combination of the two, as in Fig. 4.5 and Fig. 4.6. We can also see that for β equal to zero, $\tilde{\rho}$ is equal to $\bar{\rho}$, i.e. Eq. (4.39) does nothing to linearly filtered density $\bar{\rho}$. For β approaching infinity, Eq. (4.39) acts as the step function in Eq. (4.37), i.e. only if $\bar{\rho}$ equal to η will $\tilde{\rho}$ be η, and for all other values of $\bar{\rho}$, $\tilde{\rho}$ is equal to 0 or 1. For other cases of β, Eq. (4.39) acts as a penalty forcing the intermediate density $\bar{\rho}$ moving towards 0 and 1. In practical optimization process, β is set to a small value (e.g., 0.1) at first, and increased gradually during iteration.

By properly choosing the parameter η, the following relation in the volume preserving Heaviside filter should be satisfied, i.e.,

$$\sum_{i=1}^{N} \bar{\rho} V_i = \sum_{i=1}^{N} \tilde{\rho} V_i \tag{4.40}$$

where N is the total number of elements and V_i is the volume of the ith element. Note that Eq. (4.40) has only one unknown number η, so η can easily be determined by solving Eq. (4.40) using a simple one-dimensional search (e.g. Bi-section method or golden section search method).

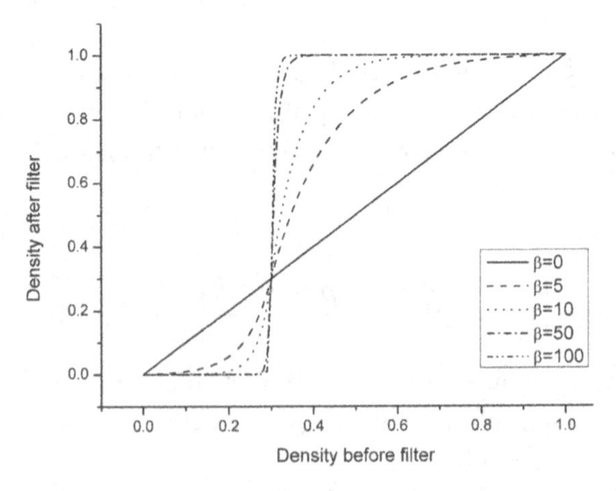

Fig. 4.5. Smoothed volume preserving nonlinear density function with different β (η=0.3)

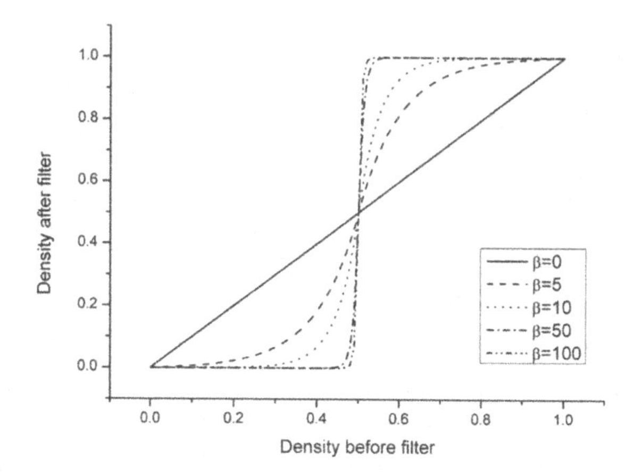

Fig. 4.6. Smoothed volume preserving nonlinear density function with different β (η=0.5)

However, due to the new parameter η is introduced, so some special attentions should be paid on the sensitivity analysis. The sensitivity of $\tilde{\rho}$ with respect to $\bar{\rho}$ is expressed as

$$\frac{\partial \tilde{\rho}_i}{\partial \bar{\rho}_i} = \left.\frac{\partial \tilde{\rho}_i}{\partial \bar{\rho}_i}\right|_{\text{ext}} + \frac{\partial \tilde{\rho}_i}{\partial \eta}\frac{\partial \eta}{\partial \bar{\rho}_i} \tag{4.41}$$

where the subscript "ext" denotes the explicit derivative of $\tilde{\rho}$ with respect to $\bar{\rho}$, i.e., it means that η is fixed. For the second term, making use of volume preserving condition shown in Eq. (4.40) leads to

$$\frac{\partial \eta}{\partial \bar{\rho}_i} = \frac{\left(1 - \frac{\partial \tilde{\rho}_i}{\partial \bar{\rho}_i}\right) V_i}{\sum_{j=1}^{\mathrm{N}} \frac{\partial \tilde{\rho}_i}{\partial \eta} V_i} \tag{4.42}$$

4.2 *Formulation for bending vibration of perforated plate*

Perforated plate, a light-weight structure, is widely used in industrial applications, which is formed by planar periodic repetition of a homogeneous 3D microstructure, but the microstructure doesn't vary along the plate thickness direction, as shown in Fig. 4.7.

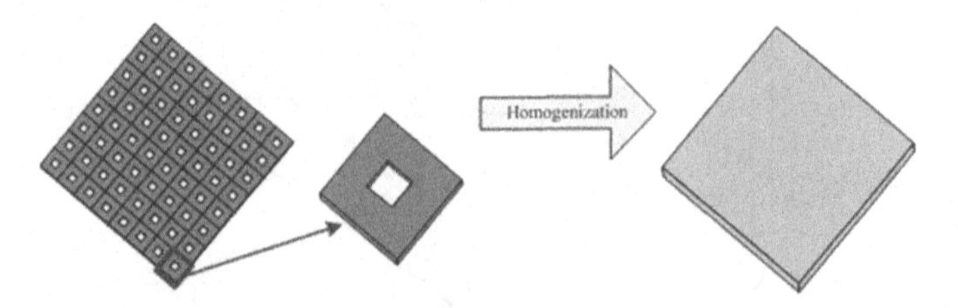

Fig. 4.7. AH based analysis of a perforated plate

Considering the bending vibration behavior of a thin perforated plate undergoing small deformations, it is reasonable to approximate the perforated plate as a classical anisotropic plate based on the effective properties resulting from the microstructure. The effective properties of this perforated plate can be obtained similarly based on mathematical AH method [142].

The general constitutive law of the perforated plate is

$$\boldsymbol{M} = \boldsymbol{D}^{\mathrm{PL}}\boldsymbol{\kappa} \tag{4.43}$$

or in the matrix form

$$\begin{aligned}\begin{Bmatrix} M_x \\ M_y \\ M_{xy} \end{Bmatrix} &= \boldsymbol{D}^{\mathrm{PL}} \begin{Bmatrix} -\dfrac{\partial^2 w}{\partial x^2} \\ -\dfrac{\partial^2 w}{\partial y^2} \\ -2\dfrac{\partial^2 w}{\partial x \partial y} \end{Bmatrix} = \frac{t^3}{12}\boldsymbol{D}^{\mathrm{H}} \begin{Bmatrix} -\dfrac{\partial^2 w}{\partial x^2} \\ -\dfrac{\partial^2 w}{\partial y^2} \\ -2\dfrac{\partial^2 w}{\partial x \partial y} \end{Bmatrix} \\ &= \frac{t^3}{12}\begin{bmatrix} D_{11}^{\mathrm{H}} & D_{12}^{\mathrm{H}} & D_{13}^{\mathrm{H}} \\ D_{21}^{\mathrm{H}} & D_{22}^{\mathrm{H}} & D_{23}^{\mathrm{H}} \\ D_{31}^{\mathrm{H}} & D_{32}^{\mathrm{H}} & D_{33}^{\mathrm{H}} \end{bmatrix} \begin{Bmatrix} -\dfrac{\partial^2 w}{\partial x^2} \\ -\dfrac{\partial^2 w}{\partial y^2} \\ -2\dfrac{\partial^2 w}{\partial x \partial y} \end{Bmatrix}\end{aligned} \tag{4.44}$$

where t is the thickness of the thin perforated plate, M_x and M_y are normal bending moments and M_{xy} is twisting moment, $\boldsymbol{\kappa}$ is curvature of the thin plate, and ω is the plate deflection in the direction of the z-axis while the z-direction coincides with the thickness direction of the thin plate and the plate is parallel to the xy plane.

The equivalent properties $\boldsymbol{D}^{\mathrm{H}}$ of perforated plate can be computed in a similar way as above plane problem [143]. The global stiffness matrix $\boldsymbol{K}$ and mass matrix $\boldsymbol{M}$ for perforated plate can be calculated similarly. $\boldsymbol{B}^{\mathrm{PL}}$ and $\boldsymbol{N}^{\mathrm{PL}}$ are the curvature-displacement matrix and the shape function matrix of plate element at the macro scale, respectively.

$$\boldsymbol{K} = \sum_{i=1}^{\mathrm{NE}} f(P_i) \int_{\Omega^{\mathrm{e}}} \left(\boldsymbol{B}^{\mathrm{PL}}\right)^{\mathrm{T}} \times \boldsymbol{D}^{\mathrm{PL}} \times \boldsymbol{B}^{\mathrm{PL}} \mathrm{d}\Omega^{\mathrm{e}} = \sum_{i=1}^{\mathrm{NE}} f(P_i) \boldsymbol{K}_i^* \tag{4.45}$$

$$\boldsymbol{M} = \sum_{i=1}^{\mathrm{NE}} \left(P_i \times \int_{\Omega^{\mathrm{e}}} \eta \times \rho^{\mathrm{PAM}} \left(\boldsymbol{N}^{\mathrm{PL}}\right)^{\mathrm{T}} \boldsymbol{N}^{PL} \mathrm{d}\Omega^{\mathrm{e}} \right) = \sum_{i=1}^{\mathrm{NE}} P_i \times \boldsymbol{M}_i^* \tag{4.46}$$

4.3 *Numerical examples*

Several numerical examples, including plane and perforated plate problems and multi domain design problems, are given in order to validate the proposed two-scale optimization method. The base material is isotropic with Young's modulus $\mathrm{E} = 7 \times 10^{10}\mathrm{Pa}$, Poisson's ratio $\upsilon = 0.3$, and mass density $\vartheta = 2700\mathrm{kg/m}^3$. The mesh is 25×25 for the microstructure design domain (Eight-node bilinear plane element).

For suppressing the numerical instability, the Heaviside density filtering technique together with continuation method is adopted in the following numerical examples. For the two-scale optimization, the macro and micro design variables are defined in different design domains, namely macro and micro design domains. Through numerical experiments, it is recommended that different parameters β^{MA} and β^{MI} respectively for macro and micro design variables are chosen. Thus, the macro and micro densities after using the Heaviside density filtering technique become

$$\tilde{P} = \begin{cases} \eta^{\mathrm{MA}} \left[e^{-\beta^{\mathrm{MA}}\left(1-\frac{\bar{P}}{\eta^{\mathrm{MA}}}\right)} - \left(1 - \frac{\bar{P}}{\eta^{\mathrm{MA}}}\right) e^{-\beta^{\mathrm{MA}}} \right] & 0 \le \bar{P} \le \eta^{\mathrm{MA}} \\ (1-\eta^{\mathrm{MA}}) \left[1 - e^{-\beta^{\mathrm{MA}}\left(\frac{\bar{P}-\eta^{\mathrm{MA}}}{1-\eta^{\mathrm{MA}}}\right)} + \frac{(\bar{P}-\eta^{\mathrm{MA}}) e^{-\beta^{\mathrm{MA}}}}{1-\eta^{\mathrm{MA}}} \right] + \eta^{\mathrm{MA}} & \eta^{\mathrm{MA}} < \bar{P} \le 1 \end{cases} \tag{4.47}$$

$$\tilde{\rho} = \begin{cases} \eta^{\mathrm{MI}}\left[e^{-\beta^{\mathrm{MI}}\left(1-\frac{\bar{\rho}}{\eta^{\mathrm{MI}}}\right)} - \left(1-\frac{\bar{\rho}}{\eta^{\mathrm{MI}}}\right)e^{-\beta^{\mathrm{MI}}}\right] & 0 \le \bar{\rho} \le \eta^{\mathrm{MI}} \\ (1-\eta^{\mathrm{MI}})\left[1-e^{-\beta^{\mathrm{MA}}\left(\frac{\bar{\rho}-\eta^{\mathrm{MI}}}{1-\eta^{\mathrm{MI}}}\right)} + \frac{(\bar{\rho}-\eta^{\mathrm{MI}})e^{-\beta^{\mathrm{MI}}}}{1-\eta^{\mathrm{MI}}}\right] + \eta^{\mathrm{MI}} & \eta^{\mathrm{MI}} < \bar{\rho} \le 1 \end{cases} \tag{4.48}$$

The Heaviside density filter is employed using a continuation approach where low values of β^{MA} and β^{MI} are used in the start, e.g., $\beta^{\mathrm{MA}} = 0$ and $\beta^{\mathrm{MI}} = 0$. Then their values are gradually increased until the satisfactory results are obtained, if lots of intermediate densities exist in the converged topology. However, when the two parameters β^{MA} and β^{MI} are increased uniformly clear topologies of macro and micro structures are hard to be obtained simultaneously. Through numerical experiments, it is found that convergence is relatively stable when the two parameters β^{MA} and β^{MI} are changed independently, e.g., $\beta^{\mathrm{MA}} = \beta^{\mathrm{MA}} + 1$ and $\beta^{\mathrm{MI}} = \beta^{\mathrm{MI}} + 4$ after one converged optimization. Meanwhile, in the Heaviside density filter using the continuation method it is normally recommended to start with large filter radiuses and gradually decrease them. In our numerical experiments, the initial values of the radiuses R and r are respectively about three or four times larger than the lengths of the macro and micro elements, and the radiuses R and r approximately equal the lengths of the macro and micro elements respectively when two-scale optimization is finished.

Numerical example 1

The first example is topological design of a beam-like macrostructure modeled by 2D plane elements. As shown in Fig. 4.8, the admissible macro design domain is a $80.0\mathrm{m} \times 50.0\mathrm{m}$ rectangle with clamped support at the left wall and a concentrated mass $\mathrm{M}_0 = 216000.0\mathrm{kg}$ at the center of the right side. Finite element model of 48×30 eight-node bilinear plane elements are utilized for dynamic analysis and optimum design. The design objective is to maximize the fundamental eigenfrequency for different prescribed base material volume fraction and different prescribed micro material volume fraction. Table 4.1 lists topological designs of macro and micro structures with variation of available base material at a specific micro volume fraction $\overline{\varsigma^{\mathrm{MI}}} = 40\%$. Table 4.1 lists topological designs of macro and micro structures with varying specific micro volume fraction for available base material amount $\bar{\varsigma} = 20\%$.

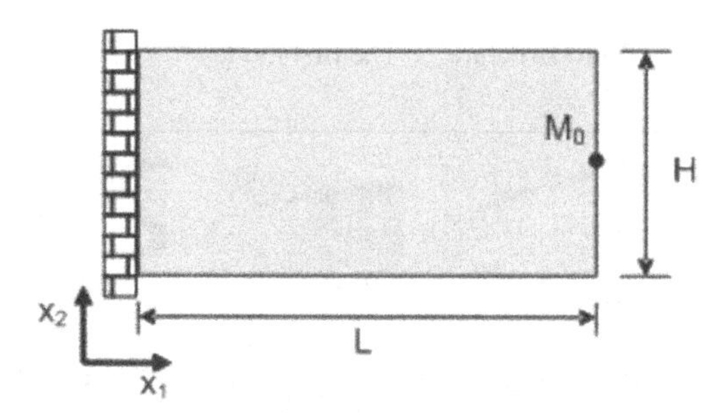

Fig. 4.8. Admissible design domain of Beam-like 2D structure clamped at the left wall

The initial values for macro design variable are uniform with value 0.5. The initial value for micro design variable is proportional to its distance from the center of the unit cell domain.

It can be seen from Table 4.1 that for specific constant micro volume fraction 40%, the fundamental frequency of the macro structure rises, and the configurations of macro and micro structures change correspondingly with the increase of available base material. The configurations of the macro structure change from 2-bar like structure to more complicated structure when the base material increases from 5% to 25%. For the base material amount 25% the material microstructure is like Kagome cell, which is believed a very good microstructure. Iteration histories of the first three eigenfrequencies for the case of the fourth row in Table 4.1 are given in Fig. 4.9, which shows the fundamental eigenfrequency always remains unimodal during optimization process for this case. Moreover, we checked that the optimized fundamental eigenfrequency is unimodal and differentiable for all test cases presented in this section.

Table 4.1. Topological designs of macro and micro structures for varying available base material at specific micro volume fraction 40%

$\bar{\varsigma}$ (%)	λ_1	Macro structure	The first mode	Micro structure	
				Cell	4×4 array
5	67.39		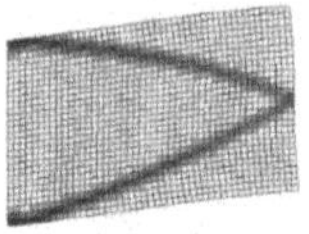		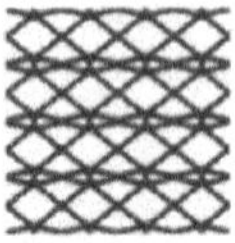
10	156.81	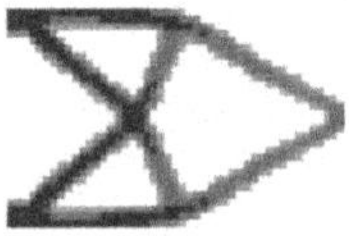	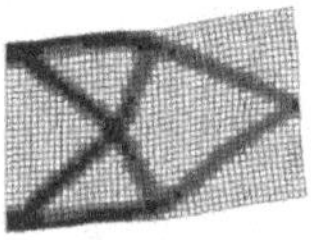		
20	487.99				
25	593.81	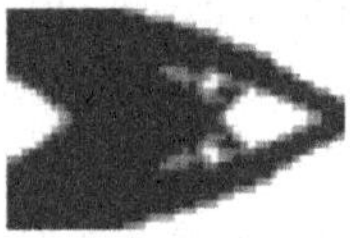			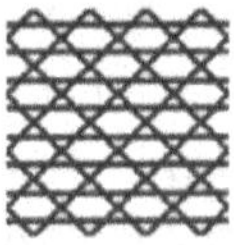

In Table 4.2 the six columns list the specific micro volume fraction, the fundamental frequency of optimum macro structure, its topology and fundamental vibration mode, the optimum topology of the unit cell and the optimum material microstructure. For all examples, the amount of available base material is 20%. Because the amount of available base material is fixed, the volume fraction of cellular material in macro design domain decreases as the specific micro volume fraction of base material at the micro scale increases. It can be seen that the fundamental eigenfrequency increases as the specific micro volume fraction of base material at the micro scale increases. In the extreme case when the specific micro volume fraction is 100%, which means that solid material without porosity is used to construct the structure, the two-scale optimization degenerates to the traditional macro structure topology optimization. This extreme case gives the highest fundamental frequency. This interesting observation somehow is unexpected because it is often claimed in literatures that ultralight material such as truss-like material has high stiffness-weight ratio. However, structures made of porous material usually undertake other functions such as active cooling, noise damping and thermal insulation. Though the present two-scale optimization method gives optimum topology of macro structure and micro structure simultaneously with the objective of maximum fundamental frequency, this optimum ultralight structure provides a good initial configuration for further considering multifunctional applications.

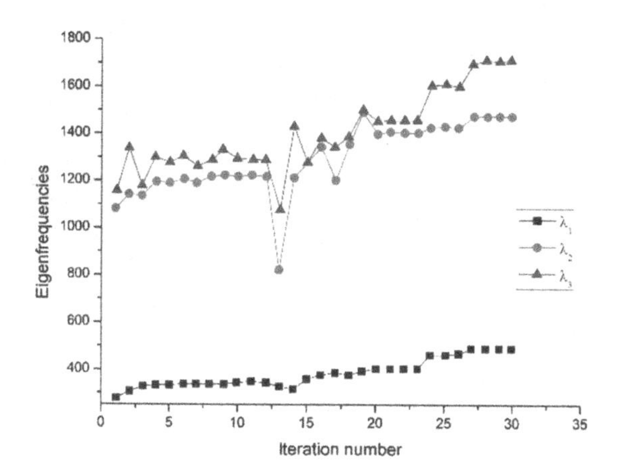

Fig. 4.9. Iteration histories of the first three eigenfrequencies for the process leading to the topology designs in fourth row of Table 4.1 with available base material amount $\bar{\varsigma} = 20\%$ and specific micro volume fraction $\overline{\varsigma^{MI}} = 40\%$

Table 4.2. Topological designs of macro and micro structures with varying micro volume at available base material amount 20%

$\overline{\varsigma^{MI}}$ (%)	λ_1	Macro structure	The first mode	Micro structure	
				Cell	4×4 array
30	435.81				
40	487.99				
60	508.96				
100	558.02				

Numerical example 2

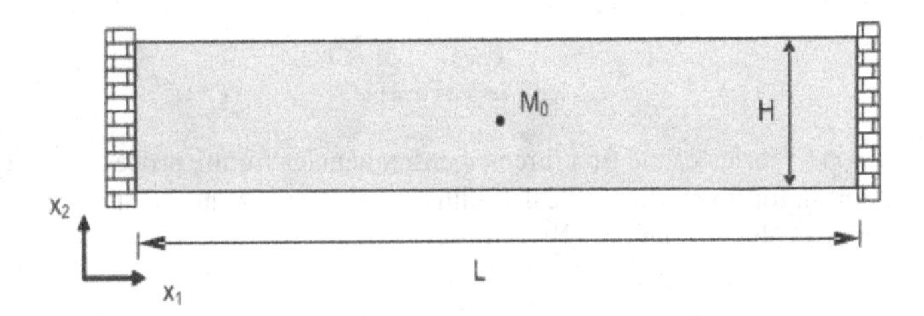

Fig. 4.10. Admissible design domain of a beam-like 2D structure with two clamped ends

In the second example as shown in Fig. 4.10, the admissible macro design domain is a $14.0\ \mathrm{m} \times 2.0\ \mathrm{m}$ rectangle with fixed support boundaries at the left and right sides and a concentrated mass $M_0 = 1512.0\ \mathrm{kg}$ at the center of the design domain. Finite element models of 140×20 eight-node plane elements are utilized. The design objective is to maximize the fundamental eigenfrequency for prescribed base material volume fraction $\overline{\varsigma} = 20\%$ and prescribed micro material volume fraction $\overline{\varsigma^{\mathrm{MI}}} = 40\%$.

Comparisons between two-scale optimization and single micro-scale optimization is then performed. In the two-scale optimization method above, macro and micro densities are introduced as independent design variables for macro structure and material microstructure. In order to illustrate the advantages of the two-scale optimization method, we compare the optimum designs from the two-scale optimization and the single micro-scale optimization. In the single micro-scale optimization, only the material microstructure needs to be designed and the macrostructure is fixed in the optimization. It means that the macro design domain is filled up with the uniform cellular material everywhere. The single microscale optimization for maximizing the fundamental frequency is formulated as

$$\begin{aligned} \text{find} \quad & \boldsymbol{X} = \{\rho\} \\ \text{obj} \quad & \max\{\min\{\lambda_j = \omega_j^2\},\ j = 1, \dots, \mathrm{J}\} \\ \text{s.t.} \quad & \boldsymbol{K}\phi_j = \lambda_j \boldsymbol{M}\phi_j, j = 1, \dots, \mathrm{J} \\ & \phi_j^T \boldsymbol{M}\phi_k = \delta_{jk}, k, j = 1, \dots, \mathrm{J} \\ & \rho^{\mathrm{PAM}} = \frac{\int_Y \rho^{\mathrm{MI}} dY}{V^{\mathrm{MI}}} = \overline{\varsigma^{\mathrm{MI}}} \\ & 0 < \underline{\rho} \le \rho_l \le \overline{\rho}, l = 1, \dots, \mathrm{n} \end{aligned} \tag{4.49}$$

where micro densities are introduced as design variables for material microstructure topology optimization. The macro densities P_i=1 (i=1, ..., NE) are fixed. Now let us apply the Eq. (4.49) to the example in Fig. 4.10. The optimized results of the two-scale and single microscale optimization are shown in Table 4.3 and Table 4.4, respectively.

Comparing the results in Table 4.3 and Table 4.4, the fundamental frequency 40822.96 obtained by two-scale optimization in Table 4.3 is higher than the fundamental frequency 36221.73 obtained by single micro-scale optimization in the fourth row of Table 4.4 for same micro volume fraction $\overline{\varsigma^{\mathrm{MI}}} = 40\%$. Furthermore, base material amount $\overline{\varsigma}$ used in two-scale optimization is 20% in Table 4.3, but 40% base material is used in single micro-scale optimization in the fourth row of Table 4.4 for same micro volume fraction $\overline{\varsigma^{MI}} = 40\%$. When the actually used base material

Table 4.3. Topological design of macro and micro structure

$\bar{\varsigma}$ (%)	$\overline{\varsigma^{MI}}$ (%)	λ_1	Macro structure	Micro structure	
				Cell	4×4 array
20	40	40822.96			
The first mode of the macro structure					

Table 4.4. Topological design of single micro-scale optimization

$\overline{\varsigma^{MI}}$ (%)	λ_1	The first mode of Macro structure	Micro structure	
			Cell	4×4 array
20	15674.20			
30	23400.65			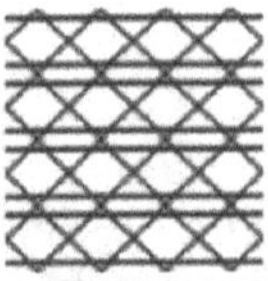
40	36221.73			

amount is also 20% for single micro-scale optimization shown in the second line of Table 4.4, the fundamental frequency is 15674.20 which is much lower than the fundamental frequency 40822.96 obtained by two-scale optimization for same amount of base material in Table 4.3. Thus, the two-scale design optimization realizes the optimized distribution of base material at macro and micro scales and obtains the optimized configurations of macro and micro structures simultaneously. It is interesting to observe that the optimum unit cell in the two-scale optimization is a Kagome cell (cf. Table 4.3), which is believed a good configuration for the future multifunctional application of cellular materials.

Numerical example 3

As shown in Fig. 4.11(a) and (b), the design domain at the macroscale is a $3.0\ \mathrm{m} \times 3.0\ \mathrm{m}$ square area with a $1.0\ \mathrm{m} \times 1.0\ \mathrm{m}$ square break in the left side and fixed at the left side. The finite element models of 3200 eight-node plate elements are utilized. The design objective is to maximize the fundamental eigenfrequency for prescribed base material volume fraction 25% and prescribed micro material volume fraction 40%. Two concentrated masses with same magnitude $\mathrm{M_0} = 432.0\ \mathrm{kg}$ are attached to different location of the design domain, as shown in Fig. 4.11(a) and (b). In Fig. 4.11(b), $\mathrm{h} = 1.05\ \mathrm{m}$.

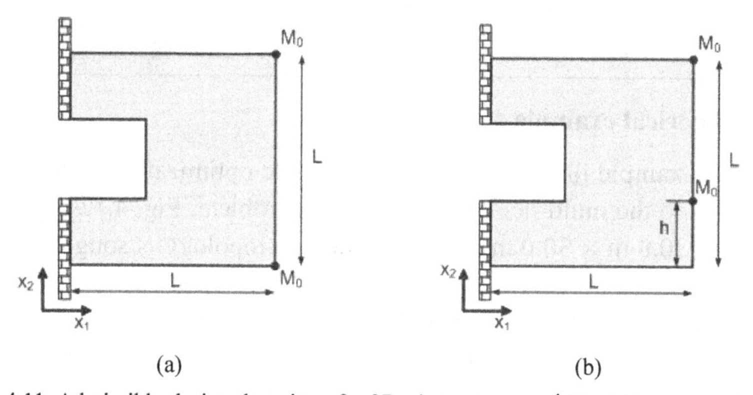

Fig. 4.11. Admissible design domains of a 2D plane structure with different locations of concentrated masses

Different locations of concentrated masses affect the optimized configurations of macro and micro structures. When the locations of two concentrated masses are symmetric shown in Fig. 4.11(a), the macro and micro structures have symmetry. When this symmetry is violated, the optimized configurations of macro and micro structures also lose the symmetry, cf. Fig. 4.11(b).

Table 4.5. Topological design of macro and micro structures considering different locations of concentrated masses

(a) the case of the Fig. 4.11(a), where $\lambda_1 = 222886.19$

$\bar{\varsigma}$ (%)	$\overline{\varsigma^{MI}}$ (%)	Macro structure	The first mode	Micro structure	
				Cell	4 × 4 array
25	40	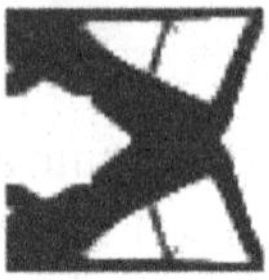	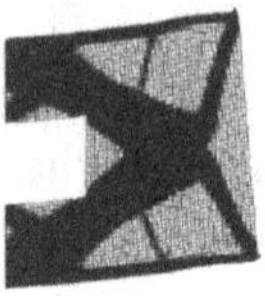	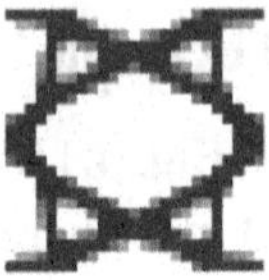	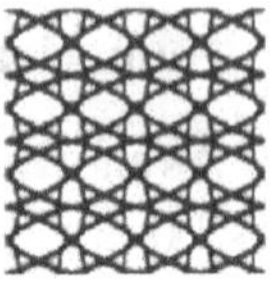

(b) the case of the Fig. 4.11(b), where $\lambda_1 = 323886.92$

$\bar{\varsigma}$ (%)	$\overline{\varsigma^{MI}}$ (%)	Macro structure	The first mode	Micro structure	
				Cell	4 × 4 array
25	40	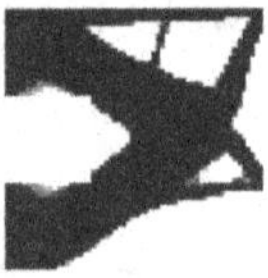	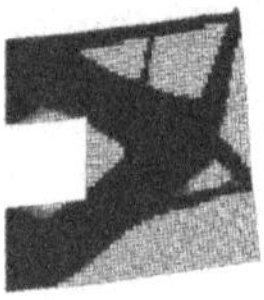	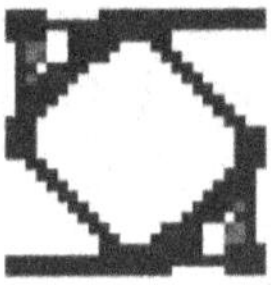	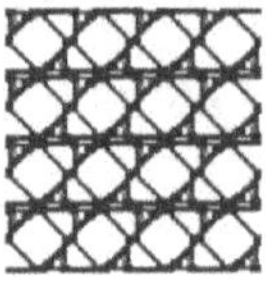

Numerical example 4

This example illustrates how the two-scale optimization method can be applied to the multi-domain optimization problem. Fig. 4.12 depicts a structure (30.0 m × 50.0 m) whose optimized topology is sought in two design domains, respectively denoted by "$Area_1$ and $Area_2$", and the structure has a bar at the center of the domain referred to as a non-design domain (30.0 m × 4.0 m). The structure needs to support three lumped masses with same magnitude $M_0 = 27000.0$ kg distributed in the right design domain "$Area_2$", as shown in Fig. 4.12. In this example, the objective is to maximize the fundamental eigenfrequency of the structure so as to limit its vibration response under certain operating conditions. It is assumed that the total amount of the available base material is 18.4% and the relative density of the microstructure is 40%.

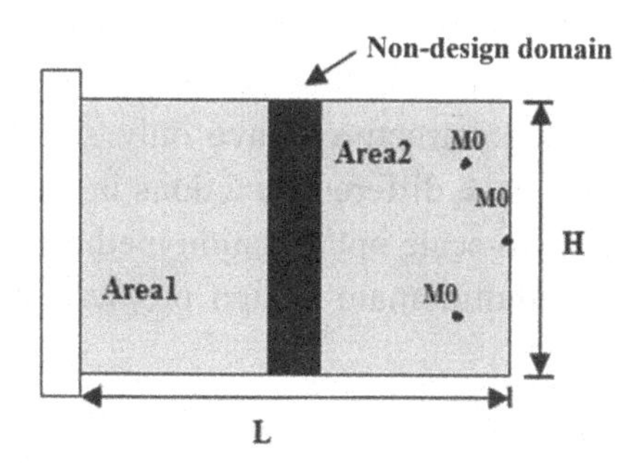

Fig. 4.12. Admissible design domain of a 2D structure with two design domains and a non-design area

Table 4.6. Topological design of macro and micro structures with multi design domains and a non-design area, where the first eigenvalues of the two cases are $\lambda_1 = 1147.84$ and $\lambda_1 = 1458.89$, respectively

$\overline{\varsigma_1}$ (%)	$\overline{\varsigma_2}$ (%)	$\overline{\varsigma^{MI}}$ (%)	Macrostructure	The first mode	Micro structure	
					Cell	4×4 array
9.2	9.2	40				
13.8	4.6	40				

Table 4.6 lists two cases for the design which have different amount of base material in two design domains. In first case, the base material of the total amount 18.4% is assigned equally to the $Area_1$ and $Area_2$. In the second case, the base material of 13.8% and 4.6% is assigned to the $Area_1$ and $Area_2$, respectively. In these two cases, the relative density of the microstructure is given as 40%. It is assumed that the structures in the two design domains and the non-design domain are made of same cellular material with uniform microstructure. When base material of total 18.4% amount is assigned to different domains, different assignments can result

in different configurations at the macro scale. However, because the constraint of microstructural relative density does not change the configurations of the microstructures have only slight variation. Thus, when different domains have different functions in the structure made of ultralight material, the two-scale optimization method presented here can be used to realize the multidomain design problem at both macro and micro scales.

Numerical example 5

Following the two-scale optimization formulation, optimized topology design of thin perforated plate at the macro scale and optimized configuration of microstructure at the micro scale can be obtained simultaneously.

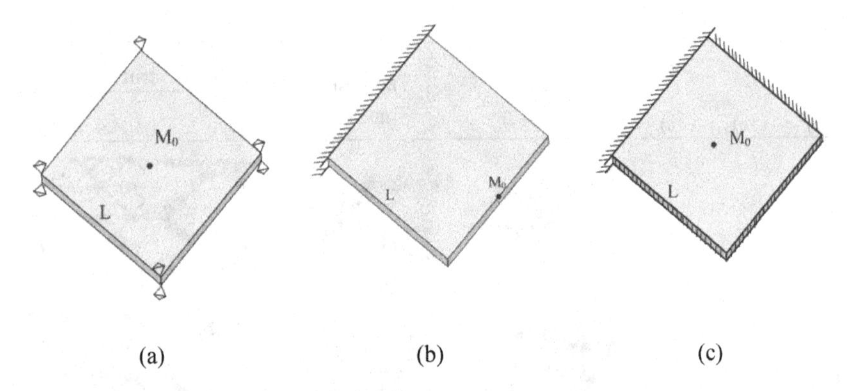

Fig. 4.13. Admissible design domains of perforated plate with different boundary conditions: (a) Simple supports at four corners and concentrated mass $M_0 = 1.8$ kg at the center of the structure. (b) One edge clamped, other edges free, and concentrated mass $M_0 = 1.8$ kg attached at the mid-point of the edge opposite to the clamped one. (c) Four edges clamped and concentrated mass $M_0 = 0.54$ kg at the center.

Three perforated plates, see Fig. 4.13(a), (b) and (c), are studied here. They have the same admissible design domain (1.0 m × 1.0 m), but three different boundary conditions and attached concentrated masses. The detailed descriptions of the boundary conditions are given in the captions of Fig. 4.13. The thickness of the plate is assumed as 0.01m.

Table 4.7 lists topological design of macro and micro structures at specific micro volume fraction $\overline{\varsigma^{MI}} = 40\%$ and available base material amount $\bar{\varsigma} = 20\%$. The resulting optimum macro and micro topologies agree with our common sense very well. It is clear in Fig. 4.14 that the fundamental eigenfrequency always remains unimodal during the optimization process.

Table 4.7. Topological design of macro and micro structures for perforated plates only with planar periodicity (a), (b), and (c)

(a) Simple supports at four corners (eigenvalue $\lambda_1 = 6483.67$)

$\bar{\varsigma}$ (%)	$\overline{\varsigma^{MI}}$ (%)	Macro structure	The first mode	Micro structure: Cell	Micro structure: 4×4 array
20	40	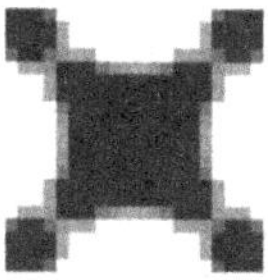	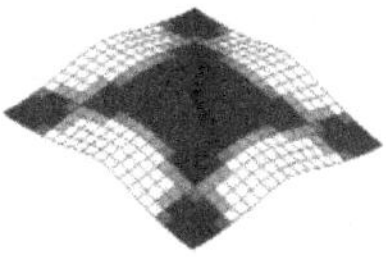	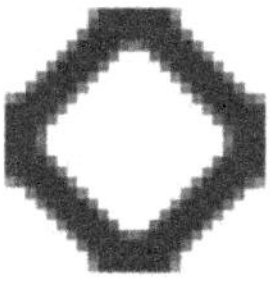	

(b) One edge clamped, other edges free (eigenvalue $\lambda_1 = 1254.48$)

$\bar{\varsigma}$ (%)	$\overline{\varsigma^{MI}}$ (%)	Macro structure	The first mode	Micro structure: Cell	Micro structure: 4×4 array
20	40		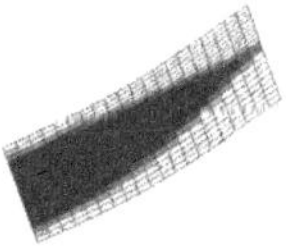		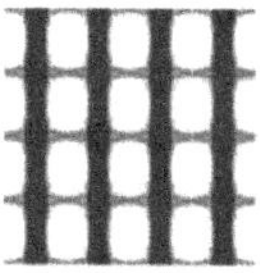

(c) Four edges clamped (eigenvalue $\lambda_1 = 63547.57$)

$\bar{\varsigma}$ (%)	$\overline{\varsigma^{MI}}$ (%)	Macro structure	The first mode	Micro structure: Cell	Micro structure: 4×4 array
20	40	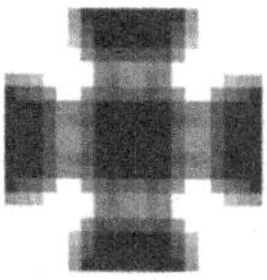	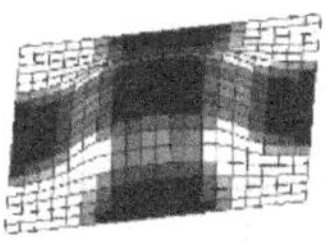		

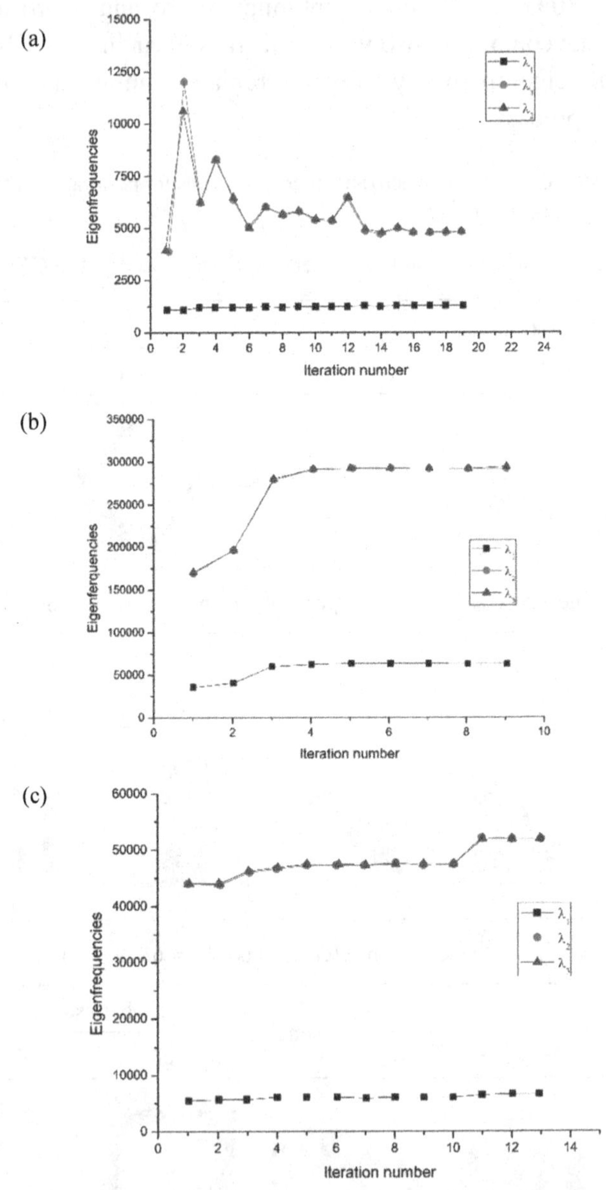

Fig. 4.14. Iteration histories of the first three eigenfrequencies for the process leading to the topology designs in Table 4.7. (a) One edge clamped, other edges free; (b) Four edges clamped; (c) Simple supports at four corners

4.4 *Concluding remarks*

In this chapter, a concurrent multiscale optimization framework based on PAMP is proposed to maximize the fundamental frequency of cellular structures. The two-scale design optimization formulation and numerical treatment for maximum frequency design are presented. It realizes the optimized distribution of base material at macro and micro scales and obtains the optimized configurations of macro and micro structures simultaneously. These novel configurations provide the guideline for further study on structure design of cellular materials in vibration environments.

The results of the two-scale topology optimization in this chapter give good mechanical properties, which offers a basis of further practical applications of the cellular structure with an excellent vibration resistance, e.g., the optimum microstructure provides a good initial configuration for the future multifunctional application of cellular materials. Additionally, different microstructures are required for different structures under different conditions when we use cellular material to construct the macrostructures. Two-scale design optimization of structural macro-topology and material micro-topology is an important tool for ultra-light structure design.

It will be an interesting and challenging work to extend this proposed optimization method to more realistic applications, e.g., including mechanical failure in constraints and multifunctional performances such as active or passive heat transfer, vibration isolation, and mechanical material failure into one system.

Furthermore, it should be pointed out that only global vibration occurs in numerical examples. The local vibration may appear when more orders of eigenmode are considered. Actually, for the optimization of the fundamental frequency of cellular structures, some opening issues still exist and need further studies, such as local vibrations, eigenmode switching, coincidence eigenvalues, singularity phenomenon of eigenfrequency constraints and so on.

Chapter 5

Multiscale Thermoelastic Optimization and the Merit of Porous Material

As a lightweight material, porous structure has been widely used in practical engineering. However, for porous structure, it is of a more promising applications in the multi-functional field. Based on this conception, the concurrent multiscale optimization based on PAMP method is applied to meet the requirement under thermoelastic environments. As aforementioned, two kinds of independent design variables are defined to describe the base material distribution at the microscale and the porous material layout at the macroscale. Here the AH method is employed to integrate the separated two-scale design variables into a complete system, and some interesting features of the two-scale coupling effect will be exhibited. In the following, two cases, i.e., single- and multi-objective optimization will be considered.

For thermoelastic optimization, two kinds of objectives are often utilized, i.e. the structural compliance and strain energy. Many literatures have given detailed discussions regarding the differences between these two objectives. Generally, structural compliance is common employed to realize maximizing stiffness design. It is logical to extend the structural compliance defined objective to the optimization of stiffness of thermoelastic structures. Meanwhile, the strain energy defined objectives have been demonstrated be an estimation of mean von-Mises stress, thus it is helpful for the stress reduction. In this chapter, the structural compliance defined objective is employed and the merits of porous material will be revealed in-depth.

For the single-objective optimization, the effect of the temperature is studied. It is found that the porous material has an inherent advantage in thermoelastic applications. In chapter 2, numerical example shows that the solid microstructure is a good choice to enhance the structural stiffness. Whereas, when mechanical and thermal loads are considered simultaneously, the porous material can better meet the complex requirements on thermoelastic field. Furthermore, the benchmark examples are given to reveal the rationality of optimized results that the varying of macro-topology as the temperature increases when structural compliance is used to define the objective function.

For the multi-objective optimization model, it attempts to find minimum structural compliance under only mechanical loads and minimum thermal expansion of the surfaces we are interested in under only thermal loads. The advantages of the concurrent optimization model to single scale topology optimization model in improving the multi-objective performances of the thermoelastic structures are investigated. The proposed multi-objective concurrent optimization model is applied to a sandwich elliptically curved shell structure, an axisymmetric structure and a 3D structure, to investigate its effectiveness. The influences of available material volume fraction and weighting coefficients are also discussed.

The work of this chapter is mainly related with references [60, 144].

5.1 *Single-objective concurrent multiscale optimization of thermoelastic structures and materials*

5.1.1 *PAMP method for multiscale optimization of thermoelastic structures*

As shown in Fig. 5.1, the PAMP model is employed for concurrent multiscale optimization of thermoelastic structures below. In this model, it is assumed that the optimal structure is made of homogeneous anisotropic porous materials. The effective elastic tensors of the porous materials can be obtained through the homogenization method based on the material distribution in the micro-scale unit cell. To be specific, we have

$$E_{ijkl}^{\mathrm{H}} = \frac{1}{|Y|}\int_Y \left[(\rho(\boldsymbol{y}))^{\alpha} E_{ijkl}^{\mathrm{S}} - (\rho(\boldsymbol{y}))^{\alpha} E_{ijst}^{\mathrm{S}} \frac{\partial \chi_s^{kl}(\boldsymbol{y})}{\partial y_t} \right] dY \tag{5.1}$$

where is α a penalization parameter introduced to suppress the intermediate density values implicitly, E_{ijkl}^{H} and E_{ijkl}^{S} $(i,j,k,l = 1,2,3)$ represent the components of the fourth-order effective elastic tensors of the porous materials and the solid base material, respectively. In Eq. (5.1), $\chi_s^{ij}(\boldsymbol{y})$ $(i,j = 1,2,3)$ is the characteristic function associated with the unit cell and $\rho(\boldsymbol{y})$ is the artificial density defined on Y characterizing the material distribution at the microscale.

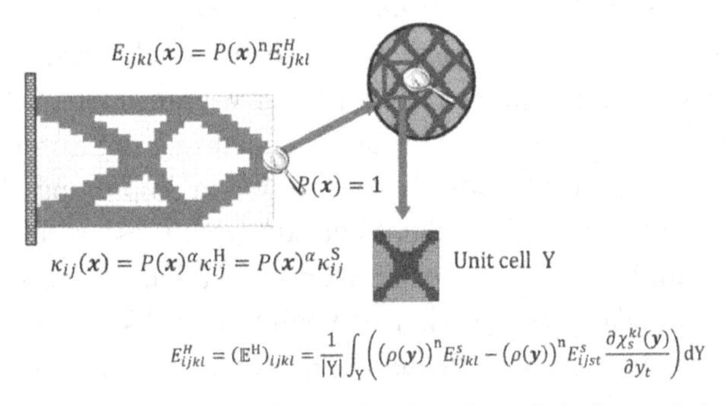

Fig. 5.1. Illustration of macro and micro design domains and the interpolation schemes in multi-scale design optimization of themoelastic structures based on PAMP optimization framework

At the macroscale, the elastic tensor at $\boldsymbol{x}$ is interpolated through the following PAMP penalty schemeß.

$$E_{ijkl}(\boldsymbol{x}) = P(\boldsymbol{x})^{\alpha} E_{ijkl}^{\mathrm{H}} \tag{5.2}$$

where $P(\boldsymbol{x})$ is the artificial density defined on the design domain characterizing the material distribution at the macroscale. Furthermore, for porous material composed of single material, the relationship $\boldsymbol{\kappa}^{\mathrm{H}} = \boldsymbol{\kappa}^{\mathrm{S}}$ also holds. Here $\boldsymbol{\kappa}^{\mathrm{H}}$ and $\boldsymbol{\kappa}^{\mathrm{S}}$ denote the effective thermal expansion tensors of the porous material and the solid base material, respectively. Accordingly, we also have

$$\kappa_{ij}(\boldsymbol{x}) = P(\boldsymbol{x})^{\alpha} \kappa_{ij}^{\mathrm{H}} = P(\boldsymbol{x})^{\alpha} \kappa_{ij}^{\mathrm{S}} \tag{5.3}$$

Therefore if $P(\boldsymbol{x})$ and $\rho(\boldsymbol{y})$ are optimized simultaneously, concurrent multiscale optimization can be achieved.

5.1.2 *Problem formulation of optimization*

In the present study, only two-dimensional problems are considered. Based on the aforementioned PAMP model for multiscale optimization, the problem considered in the present work can be formulated as follows

$$\text{find} \quad P(\boldsymbol{x}), \rho(\boldsymbol{y}) \tag{5.4}$$

$$\min \quad C = (\boldsymbol{F}_{\mathrm{m}}^{t} + \boldsymbol{F}_{\mathrm{T}}^{t})\boldsymbol{U} \tag{5.5}$$

$$\begin{aligned} \text{s.t.} \quad & \boldsymbol{KU} = \boldsymbol{F}_{\mathrm{m}} + \boldsymbol{F}_{T} \\ & \varsigma = \frac{1}{|D|}\int_D P(x)\left(\frac{1}{|Y|}\int_Y \rho(Y)dY\right)dV \leq \bar{\varsigma} \\ & 0 < P_{min} \leq P(\boldsymbol{x}) \leq 1, \forall \boldsymbol{x} \in D \\ & 0 < \rho_{min} \leq \rho(\boldsymbol{y}) \leq 1, \forall \boldsymbol{y} \in Y \end{aligned} \tag{5.6}$$

where $P(\boldsymbol{x})$ and $\rho(\boldsymbol{y})$ are the artificial density fields characterizing the material distributions in D and Y, respectively. P_{min} and ρ_{min} represent lower limit of P and ρ, respectively. The symbols ς and $\bar{\varsigma}$ are the volume fraction of available solid material and its upper limit, respectively. The superscript 't' represents the transpose of a vector or matrix. The objective function C is the work done by the mechanical load $\boldsymbol{F}_{\mathrm{m}}$ and thermal load $\boldsymbol{F}_{T}$ on the structural displacement. $\boldsymbol{K}$ is the global stiffness matrix and $\boldsymbol{U}$ is macroscopic displacement field vector. The expressions of $\boldsymbol{K}$ and $\boldsymbol{F}_{T}$ are given as

$$\boldsymbol{K} = \sum_{e} \int_{\mathrm{D_e}} \boldsymbol{B}^{\mathrm{T}}(P^{e})^{\alpha}\boldsymbol{D}^{\mathrm{H}}\boldsymbol{B}\mathrm{d}V \tag{5.7}$$

$$\boldsymbol{F}_{T} = \sum_{e} \int_{D_e} \boldsymbol{B}^{\mathrm{T}}(P^{e})^{\alpha}\boldsymbol{\sigma}_{\mathrm{T}}\Delta T\mathrm{d}V \tag{5.8}$$

where ΔT represents a temperature change experienced by the structure. P^{e} is the value of $P(\boldsymbol{x})$ on the eth element in macroscopic design domain D. $\boldsymbol{B}$ represents the strain-displacement matrix. $\boldsymbol{D}^{\mathrm{H}}$ is the effective modulus matrix. $\boldsymbol{\sigma}_{\mathrm{T}}$ is the thermal stress vector.

$$\boldsymbol{\sigma}_{\mathrm{T}} = \boldsymbol{D}^{\mathrm{H}}\boldsymbol{\kappa}^{\mathrm{H}} = \boldsymbol{D}^{\mathrm{H}}\boldsymbol{\kappa}^{\mathrm{S}} \tag{5.9}$$

where $\boldsymbol{\kappa}^{\mathrm{H}}, \boldsymbol{\kappa}^{\mathrm{S}}$ are the thermal expansion coefficients of porous material and its base material, respectively.

5.1.3 *Numerical solution aspects*

Finite element method (FEM) is used to perform structural analysis for the considered multiscale optimization problem. To this end, both macroscopic design domain D and microscopic unit cell Y are discretized by eight nodes isoparametric elements. In addition, the design variables $P(\boldsymbol{x})$ and $\rho(\boldsymbol{y})$ are also approximated by FE method and interpolated in an element-wise constant way.

(1) Sensitivity analysis

Based on Eqs. (5.4)-(5.8), the sensitivity of the objective function with respect to design variables P^e and ρ^f can be calculated as

$$\begin{aligned}\frac{\partial C}{\partial P^e} &= -\frac{\alpha}{P^e}(\boldsymbol{u}^e)^{\mathrm{t}}\left(\int_{\mathrm{D_e}} \boldsymbol{B}\,(P^e)^{\alpha}\boldsymbol{D}^{\mathrm{H}}\boldsymbol{B}\mathrm{dV}\right)\boldsymbol{u}^e \\ &\quad +2\cdot\frac{\alpha}{P^e}(\boldsymbol{u}^e)^{\mathrm{t}}\int_{\mathrm{D_e}} \boldsymbol{B}(P^e)^{\alpha}\boldsymbol{\sigma}_{\mathrm{T}}\Delta T\mathrm{d}V\end{aligned} \tag{5.10}$$

$$\begin{aligned}\frac{\partial C}{\partial \rho^f} &= -\sum_e(\boldsymbol{u}^e)^{\mathrm{t}}\left(\int_{\mathrm{D_e}} \boldsymbol{B}\,(P^e)^{\alpha}\left(\frac{\partial \boldsymbol{D}^{\mathrm{H}}}{\partial \rho^f}\right)\boldsymbol{B}\mathrm{d}V\right)\boldsymbol{u}^e \\ &\quad +2\sum_e(\boldsymbol{u}^e)^{\mathrm{t}}\int_{\mathrm{D_e}} \boldsymbol{B}(P^e)^{\alpha}\left(\frac{\partial \boldsymbol{D}^{\mathrm{H}}}{\partial \rho^f}\boldsymbol{\kappa}^{\mathrm{S}}\right)\Delta T\mathrm{d}V\end{aligned} \tag{5.11}$$

where

$$\frac{\partial \boldsymbol{D}^{\mathrm{H}}}{\partial \rho^f} = \frac{\alpha}{\rho^f}\frac{1}{|Y|}\int_{\mathrm{Y}_f}(\boldsymbol{I}-\boldsymbol{B}\cdot\boldsymbol{\chi})\left(\rho^f\right)^{\alpha}\boldsymbol{D}^{\mathrm{s}}(\mathbf{I}-\boldsymbol{B}\cdot\boldsymbol{\chi})dY \tag{5.12}$$

$\boldsymbol{u}^e$ is the displacement field of the eth element. In Eq.(5.12), $\boldsymbol{I}$ is a 3×3 identify matrix and

$$\chi = \begin{pmatrix} \chi_1^{11} & \chi_1^{22} & \chi_1^{12} \\ \chi_2^{11} & \chi_2^{22} & \chi_2^{12} \end{pmatrix}$$

is the matrix of characteristic functions associated with the unit cell. The sensitivity analysis of the volume constraint function with respect to design variables is trivial and will not be discussed here.

(2) Regularization techniques and optimization algorithm

For multiscale design and optimization, numerical problems such as the existence of grey elements in the optimized structure and the convergence of the optimization algorithms are more severe. In the present work, Heaviside function based density filtering and continuation approach (Guest et al. [102]) is employed to alleviate the numerical problems mentioned above. Furthermore, Sequential Quadratic Programming (SQP) approach is used as the optimizer to solve the optimization problem.

5.1.4 *Numerical examples*

(1) **Example 1**: A two-ends clamped beam under single mechanical load case and a uniform temperature change

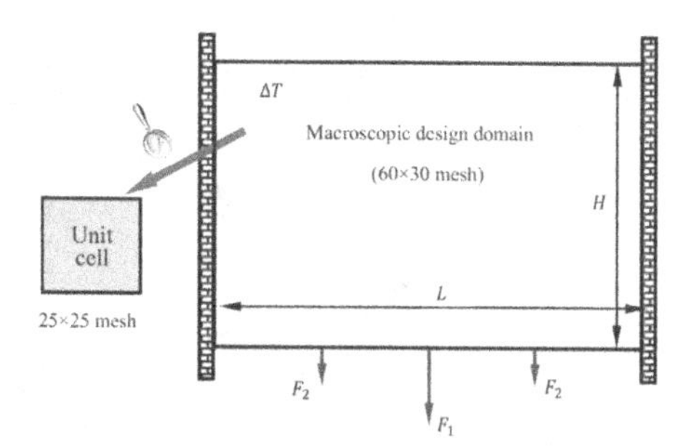

Fig. 5.2. A two ends clamped beam with single mechanical load case

In this example, a two ends clamped beam under single mechanical load case is considered. The macroscopic design domain with $L = 72$ cm and $H = 47.7$ cm is shown in Fig. 5.2. A concentrated force $F_1 = 50000$ N is applied at the central point of the lower side of the design domain and the whole structure experiences a uniform temperature change ΔT. The

Young's modulus, Poisson's ratio and the thermal expansion coefficient of the isotropic solid base material are $E^s = 100\ \text{Gpa}, \nu^s = 0.3$ and $\kappa^S = 10^{-5}$, respectively. It is intended to find the optimal topology of the macroscopic structure and the topology of the unit cell simultaneously in order to minimize the structural compliance under available solid material volume fraction constraint (i.e., $\varsigma \leq \bar{\varsigma} = 0.05$). Only half part of the design domain is considered due to the symmetry property of the problem. The resolution of the FE mesh on the macroscopic and microscopic design domains is also described in Fig. 5.2.

Table 5.1. Optimization results for example 1

ΔT (°C)	Macroscopic Optimized topology	Optimized topology of the unit cell	
		Unit Cell	4×4 arrays
0			
1			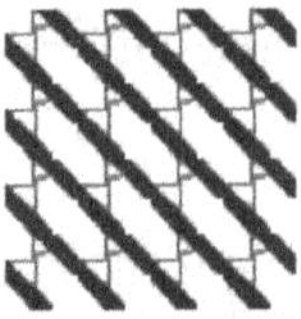
20			
40			

Table 5.1 shows optimization results with different values of ΔT. Actually, this example had been studied in Yan et al. [145] with use of the

perimeter-like constraint suggested in Zhang et al. [109] to eliminate the grey elements. In the present study, Heaviside function based density filtering and continuation approach (Guest et al. [102], Duan et al. [101]) is employed to obtain more clearer black-and-white structural topologies. Both optimized topologies of the macroscopic structure and the microscopic unit cell are shown in Table 5.1.

From the results shown in Table 5.1, it is observed that when $\Delta T = 0$, the optimization result coincides with that obtained under pure mechanical load. It is interesting to note that under this circumstance, the unit cell is filled with solid material (i.e., $\rho^{\mathrm{opt}}(\boldsymbol{y}) = 1, \forall \boldsymbol{y} \in \mathrm{Y}$) and its porosity is zero. This phenomenon has been observed in the previous chapters, which is because the sensitivity with respect to the micro design variables is larger than that of the macro as shown in Eq.(5.10)-(5.11). This implies that if the available solid material volume is fixed, it is more efficient to distribute material on the microscopic structure to enhance the structural stiffness. It is worth noting that the above result does not contradict with classical Michell truss theory since the microstructure is not allowed to vary point-wisely in the present problem.

When $\Delta T \neq 0$, it can be observed from Table 5.1 that the optimized macroscopic structure looks like a downward two-bar truss when ΔT is relatively small (i.e., mechanical load dominates) and changes to an upward three-bar truss when ΔT is relatively large (i.e., thermal load dominates). This conclusion has also been confirmed by the analytical analysis. Actually, for the upward three-bar truss like structure, the vertical components of $\boldsymbol{F}_{\mathrm{m}}$ and $\boldsymbol{F}_{\mathrm{T}}$ at the intersection point of the truss bars have different signs. This leads to a relatively small value of the structural compliance measured by the magnitude of the vertical component of the displacement vector at that point.

(2) **Example 2:** A two-end clamped beam under multiple mechanical load cases and a uniform temperature change

In this example, the above problem is re-examined by considering multiple mechanical loading cases. Here besides F_1, other two concentrated loads with magnitudes $\mathrm{F}_2 = 25000\ \mathrm{N}$ is also considered as shown in Fig. 5.2. The corresponding optimization results are presented in Table 5.2 and Table 5.3, respectively.

Table 5.2. Optimization results for example 2

ΔT (°C)	C (N · cm)	Macroscopic optimized topology	Optimized topology of the unit cell	
			Unit Cell	4 × 4 arrays
2	53651	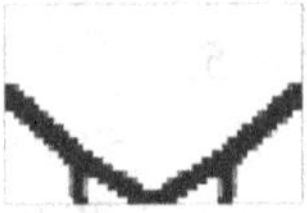		
10	63693			
30	71445			

Table 5.3. The influence of $\bar{\varsigma}$ on the forms of optimized microscopic and macroscopic structures, $\Delta T = 2$°C

$\bar{\varsigma}$	ρ^{macro}	ρ^{micro}	C (N · cm)	Macroscopic optimized topology	Optimized topology of the unit cell	
					Unit Cell	4 × 4 arrays
0.1	0.1470	0.6807	53651			
0.2	0.2121	0.9435	22566		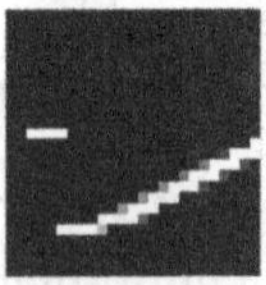	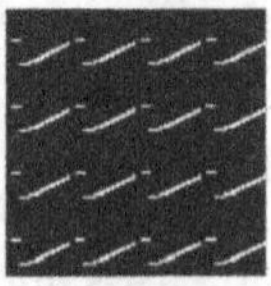
0.3	0.3204	0.9363	20041		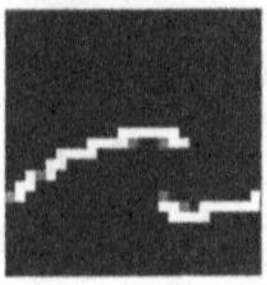	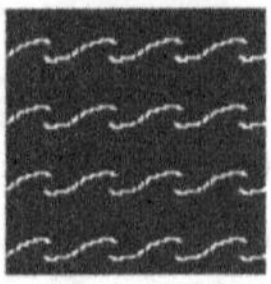

* $\rho^{\text{macro}} = \int_D P(\boldsymbol{x})dV/|D|$

Table 5.2 gives the optimization results for $\bar{\varsigma} = 0.1$. From this table, it can be observed that the value of ΔT has great influence on the forms of optimized microscopic and macroscopic structures when $\bar{\varsigma}$ is fixed. As ΔT increases, the optimized macros*copic structure transforms from a downward four-bar truss like structure to an upward arc type structure. In addition, the optimized microscopic structure changes dramatically and the form of these microstructures are quite beyond one's intuition.

Table 5.3 gives the optimization results for $\Delta T = 2°\text{C}$ under different values of $\bar{\varsigma}$. It is found from this t- able that the value of $\bar{\varsigma}$ also has great influence on the multiscale optimization results. Furthermore, even though the value of $\bar{\varsigma}$ is relatively large, the porosity of the microstructure (i.e., $\rho^{\text{micro}} = \int_{\text{Y}} \rho(y) dY / |Y|$) is always less than 1.0. This means that porous anisotropic material is indeed effective to reduce structural compliance when both mechanical and thermal loads are applied simultaneously. And the optimized configurations of the porous media shown in Table 5.1 to Table 5.3 can be regarded as open cell.

5.1.5 *Analytical validation*

Although under thermoelastic enviorment have been numerous works on multiscale design and optimization in literatures, benchmark examples that can validate the correctness (even though qualitatively) of the results obtained by these methods still lack. In this Section, it is intended to validate the multiscale concurrent optimization results obtained above by analyzing an analytically tractable example. Observing that the optimized macroscopic structures shown in Table 5.1 are very similar to truss type structures when the value of $\bar{\varsigma}$ is relatively low, this inspires us that truss model can be used to obtain some analytical benchmark solutions.

Let us consider a $\text{L} \times \text{H}$ macroscopic rectangular design domain shown in Fig. 5.3. A three-bar truss structure hinged at its left and right ends is used as a ground structure. A concentrated force F is applied at the central point of the lower side of the design domain and the whole structure experiences a uniform temperature change ΔT. It is assumed that the truss bars are composed of homogenous porous material and the Young's modulus, Poisson's ratio and thermal expansion coefficient of the solid base material are E^{S}, μ^{S} and $\kappa_{\text{T}}^{\text{S}}$, respectively. Note that $\kappa_{\text{T}}^{\text{S}}$ is also equal to the homogenized thermal expansion coefficient of the porous material (i.e., $\kappa_{\mathrm{T}}^{\mathrm{H}}$) as shown in Liu and Cheng [146], Hashin [147].

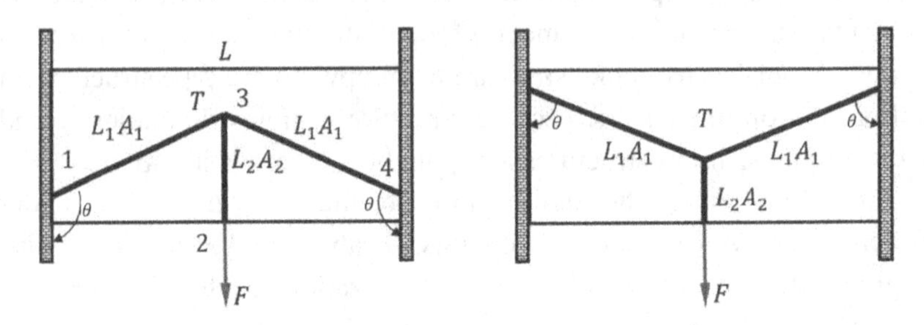

Fig. 5.3. Two candidate optimized structural configurations when H/L is relatively large

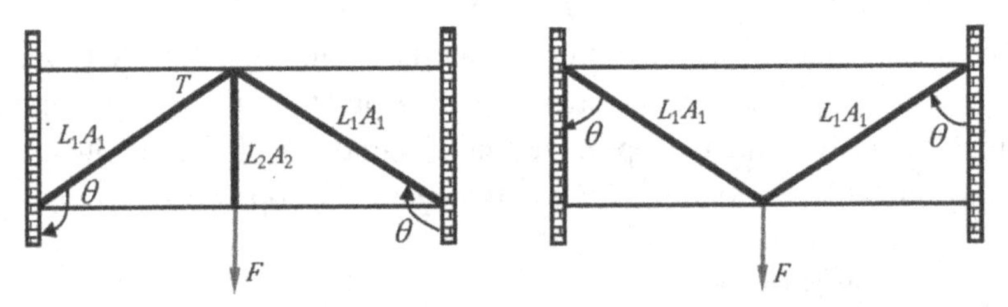

Fig. 5.4. Two candidate optimized structural configurations when H/L is small

In the following, SIMP approach is employed to find the optimal layout of the truss structure. To this end, a relationship $E^{\mathrm{H}} = \rho^{\alpha}\mathrm{E}_{\mathrm{s}}$ (ρ is the density of the porous material) between the Young's modulus of the porous material (i.e., E^{H}) and the Young's modulus of the base material (i.e., E_{s}) is assumed. Obviously, the form of the optimal structure is dependent on the ratio of $\mathrm{H/L}$. Numerical results in Table 5.1 suggest that four configurations shown in Fig. 5.3 and Fig. 5.4, respectively, can serve as the candidates of optimal solutions. Based on these observations, a set of geometry parameters, namely the length of each bar (i.e., L_1 and L_2), the cross sectional area of each bar (i.e., A_1 and A_2) and the inclined angle (i.e., θ) are also used as design variables to describe the macroscopic configurations of the truss structures.

Based on the above preparations, the compliance minimization problem under mechanical and thermal loads can be formulated as

$$
\begin{aligned}
\text{find} \quad & L_2, A_1, A_2, \theta, \rho \\
\min \quad & C \\
\text{s.t.} \quad & \rho(A_1 L/\sin\theta + A_2 L_2) \le \bar{\mathrm{V}} \\
& L_2 + \frac{L\cos\theta}{2\sin\theta} \le \mathrm{H} \\
& 0 \le L_2 \le H \;\; 0 \le \theta \le \pi
\end{aligned}
\tag{5.13}
$$

where

$$
\begin{aligned}
C = {} & \frac{\mathrm{F}^2}{E^{\mathrm{H}}}\left(\frac{\mathrm{L}}{4A_1 \sin\theta \cos^2\theta} + \frac{L_2}{A_2}\right) + 2\mathrm{F}\kappa_{\mathrm{T}}^{\mathrm{S}}\Delta T\left(\frac{\mathrm{L}}{2\sin\theta\cos\theta} + L_2\right) \\
& + \frac{E^{\mathrm{H}}\left(\kappa_{\mathrm{T}}^{\mathrm{S}}\right)^2 \Delta T^2 (A_1 \mathrm{L}/\sin\theta + A_2 L_2)}{\rho}
\end{aligned}
\tag{5.14}
$$

$\bar{V} = \bar{\varsigma}\mathrm{LH}$ and $E^{\mathrm{H}} = \rho^{\alpha}\mathrm{E}^{\mathrm{S}}$, respectively. In the above derivations, the relationship $2L_1 \sin\theta = \mathrm{L}$ has been used. In Eq. (5.13), the first constraint is the total solid base material volume constraint and the second constraint is a geometry constraint for the configuration of the truss structure. In the following, we shall discuss three loading cases separately.

(1) Only mechanical load is applied

When only mechanical load is applied, the corresponding optimization problem is

$$
\begin{aligned}
\text{find} \quad & L_2, A_1, A_2, \theta, \rho \\
\min \quad & C = \frac{\mathrm{F}^2}{E^{\mathrm{H}}}\left(\frac{\mathrm{L}}{4A_1 \sin\theta \cos^2\theta} + \frac{L_2}{A_2}\right) \\
\text{s.t.} \quad & \rho(A_1 \mathrm{L}/\sin\theta + A_2 L_2) \le \bar{V} \\
& L_2 + \frac{\mathrm{L}\cos\theta}{2\sin\theta} \le \mathrm{H} \\
& 0 \le L_2 \le \mathrm{H}; \;\; 0 \le \theta \le \pi
\end{aligned}
\tag{5.15}
$$

Since the volume constraint must be active when pure mechanical load is considered [141], it yields that $A_2 = (\bar{V}\sin\theta/\rho - \mathrm{L}A_1)/L_2 \sin\theta$. By introducing the following Augmented Lagrangian function

$$\bar{C}=\frac{\mathrm{F}^2}{E^{\mathrm{H}}}\left(\frac{\mathrm{L}}{4A_1\sin\theta\cos^2\theta}+\frac{L_2}{A_2}\right)+\lambda\left(L_2+\frac{\mathrm{L}\cos\theta}{2\sin\theta}-\mathrm{H}\right) \tag{5.16}$$
$$-\mu L_2+\tau(L_2-\mathrm{H})$$

where λ, μ and τ are Lagrange multipliers, the optimal solution of Eq. (5.15) can be found as Eq. (5.17).

$$C^{\mathrm{opt}}=\begin{cases}\dfrac{F^2L^2}{\mathrm{E^s\bar{V}}}, for\ \mathrm{L}\le 2\mathrm{H},\\ \text{with } \theta^{\mathrm{opt}}=\dfrac{\pi}{4},\rho^{\mathrm{opt}}=1,A_1^{\mathrm{opt}}=\dfrac{\bar{\mathrm{V}}}{\sqrt{2}L},\\ A_2^{\mathrm{opt}}=0,L_1^{\mathrm{opt}}=\dfrac{\sqrt{2}\mathrm{L}}{2},L_2^{\mathrm{opt}}=0,\\ \dfrac{\mathrm{F^2L^2(4H^2+L^2)^2}}{16\mathrm{E^s}\bar{V}\mathrm{H^2L^2}}, \text{for } \mathrm{L}>2\mathrm{H},\\ \text{with } \theta^{\mathrm{opt}}=\tan^{-1}\dfrac{\mathrm{L}}{2\mathrm{H}},\rho^{\mathrm{opt}}=1,A_1^{\mathrm{opt}}=\dfrac{\bar{V}}{\sqrt{\mathrm{L^2+4H^2}}},\\ A_2^{\mathrm{opt}}=0,L_1^{\mathrm{opt}}=\dfrac{\sqrt{\mathrm{L^2+4H^2}}}{2},L_2^{\mathrm{opt}}=0.\end{cases} \tag{5.17}$$

It is worth noting that the numerical results presented in Table 5.1 obtained in the continuum setting indicate that when $\mathrm{L}=72$ cm and $\mathrm{H}=47.7$ cm ($\mathrm{L}<2\mathrm{H}$), the corresponding optimal value of the inclined angle is $\theta_{\mathrm{con}}^{\mathrm{opt}}\approx 0.2489\pi$, which is in good agreement with the analytical result (i.e., $\theta^{\mathrm{opt}}=\pi/4$) shown in Eq. (5.17). Fig. 5.5 plots the optimal configuration of the truss structure under this case.

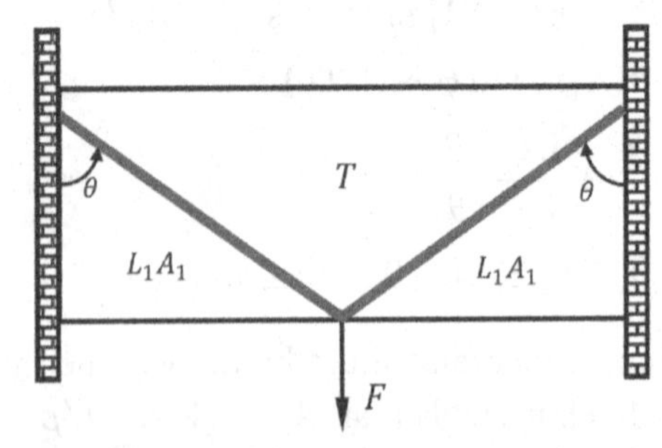

Fig. 5.5. Optimal configuration of the truss structure under pure mechanical load

(2) Only thermal load is applied

When only thermal load is applied, the corresponding optimization problem is

$$\begin{aligned}
&\text{find} \quad L_2, A_1, A_2, \theta, \rho \\
&\min \quad C = E^{\mathrm{H}} \kappa_{\mathrm{T}}^{\mathrm{S}} (\Delta T)^2 (2A_1 L_1 + A_2 L_2)/\rho \\
&\text{s.t.} \quad \rho(A_1 \mathrm{L}/\sin\theta + A_2 L_2) \le \bar{V} \\
&\qquad L_2 + \frac{\mathrm{L}\cos\theta}{2\sin\theta} \le \mathrm{H} \\
&\qquad 0 \le L_2 \le \mathrm{H}; \quad 0 \le \theta \le \pi
\end{aligned} \tag{5.18}$$

where $E^{\mathrm{H}} = \rho^{\alpha} \mathrm{E}^{\mathrm{S}}$.

For this problem, if the volume constraint must be active at the optimal solution, then it yields that $C = \mathrm{E}^{\mathrm{S}} (\kappa_{\mathrm{T}}^{\mathrm{S}})^2 \Delta T^2 \bar{V} \rho^{\alpha-1}$. This indicates that the optimal value of the objective function is independent on the values of geometry parameters (e.g., L and H) and it is obviously that the optimal solution corresponds to a degenerate solution $\rho^{\mathrm{opt}} = 0$ if the volume constaint can be inactive. This is because no material implies no thermal load.

(3) Both mechanical and thermal loads are applied

When both mechanical and thermal loads are considered, it is difficult to obtain the analytical solution of Eq. (5.17) explicitly. Therefore, SQP method is used to find the optimal solutions corresponding to two truss configurations (a three-bar truss and a two-bar truss), respectively. For comparison purpose, the same values of the geometry, material and loading data (i.e., $\mathrm{L}, \mathrm{H}, \mathrm{E}^{\mathrm{S}}, \mathrm{F}$, etc) as those in the numerical examples are used. Unless otherwise stated, a relatively small value of $\bar{\varsigma}$ (i.e., $\bar{\varsigma} = 0.05$) is adopted here in order to guarantee that the optimized structures obtained in the continuum setting are truss-like ones. Fig. 5.6 to Fig. 5.9 show the optimization results obtained under different conditions.

Fig. 5.6 plots the variation of optimal structural compliance with respect to temperature change ΔT for different truss configurations. From Fig. 5.6, it can be observed that the compliance of the three-bar truss structure (square dot line) is much less than that of the two-bar truss structure (circle dot line) for a large range of the value of ΔT (i.e., $\Delta T >$ 1°C). This means that when the value of thermal load is relatively large, compared to the two-bar truss structure, the three-bar truss structure is more appropriate to reduce the structural compliance. This result is quite consistent with the result (e.g., the transition point between the two optimal configurations) shown in Table 5.1 obtained under the continuum setting.

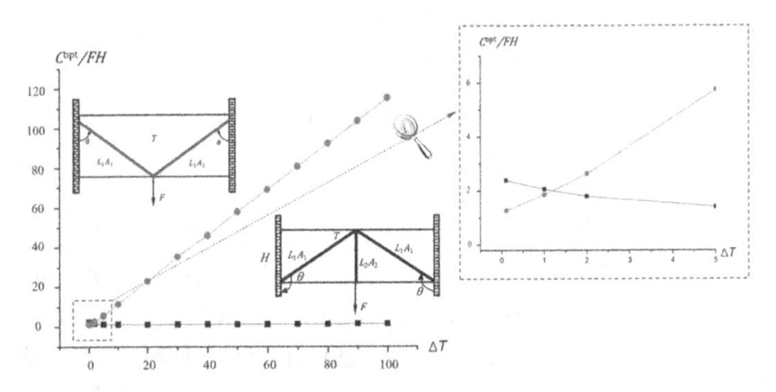

Fig. 5.6. The variation of optimal structural compliance with respect to ΔT for different truss configurations

Fig. 5.7 plots the variation of the optimal density of porous material with respect to temperature change ΔT for different truss configurations. The results shown in this figure indicate that when mechanical and thermal loads are applied simultaneously, for both two-bar and three-bar truss configurations, porous material (i.e., $\rho < 1$) is a better choice to reduce the structural compliance. The larger the value of ΔT, the smaller the optimal value of ρ. This is quite different from the pure mechanical load case where solid material ($\rho = 1$) is always better than porous material to enhance structural stiffness. The numerical results shown here also coincide with that obtained in the continuum setting.

Fig. 5.8 plots the variation of optimal inclined angle with respect to temperature change ΔT for different truss configurations. From this figure., it is found that for the two-bar truss configuration, α^{opt} is a constant value (i.e., $\theta^{\text{opt}} = \pi/4$) for all values of ΔT. For the three-bar truss configuration, when only mechanical load is considered (i.e., $\Delta T =$ 0), it can be calculated that $\theta^{\text{opt}} \cong 0.69\pi$. As ΔT increases, the optimal three-bar structure becomes more and more flat. This tendency is also exactly the same as that under the continuum setting.

Fig. 5.9 plots the variation of optimal structural compliance with respect to $\bar{\varsigma}$ for different truss configurations. An obvious trend reflects from this figure is that the optimal value of structural compliance (i.e., C^{opt}) decreases as $\bar{\varsigma}$ increases. When the value of $\bar{\varsigma}$ is relatively small, the two-bar truss configuration is better than the three-bar truss configuration in terms of C^{opt}. However, when $\bar{\varsigma} > 0.1$, the three-bar truss configuration becomes better. It is also observed when $\bar{\varsigma}$ is larger than a critical value (i.e., $\bar{\varsigma}_{\mathrm{cr}} = 0.2$ for the two-bar truss configuration and $\bar{\varsigma}_{\mathrm{cr}} = 0.35$ for the three-bar-truss configuration), the value of C^{opt} is saturated and reaches a stable value. These trends are also in accordance with those observed under the continuum setting.

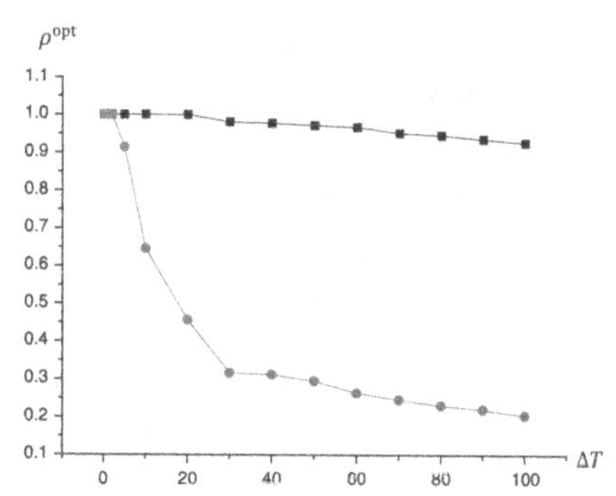

Fig. 5.7. The variation of the optimal density of porous material with respect to $\Delta\mathrm{T}$ for different truss configurations

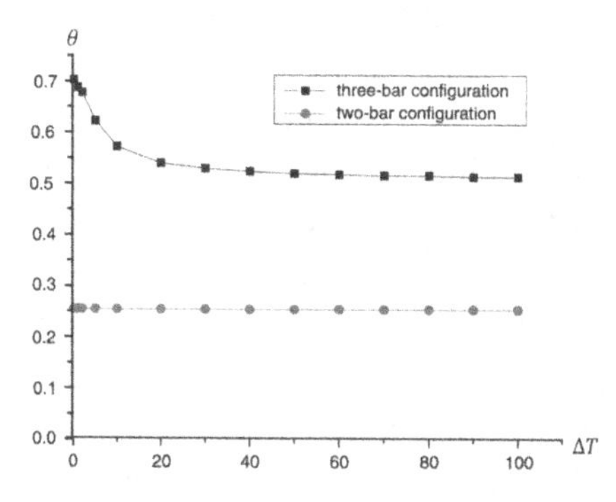

Fig. 5.8. The variation of optimal inclined angle with respect to ΔT for different truss configurations

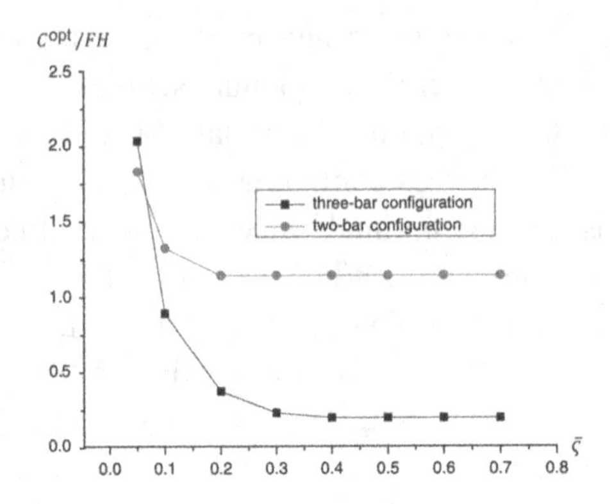

Fig. 5.9. The variation of optimal structural compliance with respect to $\bar{\varsigma}$ for different truss configurations

5.2 *Multi-objective concurrent topology optimization of thermoelastic structures and materials*

5.2.1 *Optimization formulation*

Many practical thermoelastic structures, for example space antennas and supports of space camera, experience large temperature variations that lead to extreme thermal deformation and deteriorate their functional accuracy. Therefore, thermoelastic structures should not deform too much, at least not too much in the domain where we care about. What's more, most thermoelastic structures also need to have enough stiffness or rigidity to carry mechanical loads. In this section, we design structures combining low thermal expansion in a predefined domain with high stiffness using the concurrent method. We concurrently design the topology of thermoelastic structures composed of homogeneous porous material and the microstructure of the porous material to improve the multi-objective performance of the thermoelastic structures. It is worth mentioning that if we care about thermal stress of the structure, the multi-objective optimization model should also minimize thermal stress of the structure, which will be the focus of our future work.

The thermal deformation of the structural domain concerned can be measured through

$$C_{\mathrm{T}} = \sum_{i=1}^{\mathrm{S}} (U_{\mathrm{T}}^{\lambda(i)})^2 \tag{5.19}$$

where, S is the number of nodes located in the domain, λ_i $(i = 1, 2, \ldots, \mathrm{S})$ is the concerned nodal degrees of freedom caused by the thermal load and $\boldsymbol{U}_{\mathrm{T}}$ is the nodal displacement vector under only thermal loads. The stiffness of the structure can be reflected by its structural compliance as

$$C_{\mathrm{m}} = (\boldsymbol{F}_{\mathrm{m}})^{\mathrm{t}} \boldsymbol{U}_{\mathrm{m}} \tag{5.20}$$

where $\boldsymbol{U}_{\mathrm{m}}$ is the nodal displacement vector under nodal mechanical load vector $\boldsymbol{F}_{\mathrm{m}}$.

The formulation of this multi-objective topology optimization problem can be stated as follows. Here, a simple but effective scheme, i.e., the weighted summation scheme is used to integrate the thermal deformation and structural stiffness measures into a system, i.e.,

$$\text{find} \quad \boldsymbol{x} = \{\boldsymbol{P}, \boldsymbol{\rho}\} \tag{5.21}$$

$$\text{min} \quad obj = w_{\mathrm{c}} f_{\mathrm{c}} + w_{\mathrm{u}} f_{\mathrm{u}} \tag{5.22}$$

$$\text{s.t.} \quad \boldsymbol{K}(\boldsymbol{x}) \boldsymbol{U}_{\mathrm{m}} = \boldsymbol{F}_{\mathrm{m}} \tag{5.23}$$

$$\boldsymbol{K}(\boldsymbol{x}) \boldsymbol{U}_{\mathrm{T}} = \boldsymbol{F}_{\mathrm{T}}(\boldsymbol{x}) \tag{5.24}$$

$$V = \rho^{\mathrm{PAM}} \cdot \rho^{\mathrm{MA}} \leq \bar{V} \tag{5.25}$$

$$0 \leq \delta \leq P_i \leq 1, (i = 1,2, \ldots, \mathrm{N}) \tag{5.26}$$

$$0 \leq \delta \leq \rho_i \leq 1, (i = 1,2, \ldots, \mathrm{n}) \tag{5.27}$$

where

$$\begin{aligned} f_{\mathrm{c}} &= \frac{C_{\mathrm{c}}}{C_{\mathrm{c}}^0} = \frac{(\boldsymbol{F}_{\mathrm{m}})^{\mathrm{t}} \boldsymbol{U}_{\mathrm{m}}}{(\boldsymbol{F}_{\mathrm{m}})^{\mathrm{t}} (\boldsymbol{U}_{\mathrm{m}})_0} \\ f_{\mathrm{u}} &= \frac{C_{\mathrm{u}}}{C_{\mathrm{u}}^0} = \frac{\sum_{i=1}^{\mathrm{S}} (\boldsymbol{U}_{\mathrm{T}}^{\lambda(i)})^2}{\sum_{i=1}^{\mathrm{S}} ((\boldsymbol{U}_{\mathrm{T}}^{\lambda(i)})_0)^2} \end{aligned} \tag{5.28}$$

where $C_{\mathrm{c}}^0 = (\boldsymbol{F}_{\mathrm{m}})^{\mathrm{t}} (\boldsymbol{U}_{\mathrm{m}})_0$ and $C_{\mathrm{u}}^0 = \sum_{i=1}^{\mathrm{S}} ((\boldsymbol{U}_{\mathrm{T}}^{\lambda(i)})_0)^2$ are the structural compliance and thermal expansion measure of the initial design, respectively. $\boldsymbol{F}_{\mathrm{m}}$ and $\boldsymbol{F}_{\mathrm{T}}$ are the nodal load vectors formed by the mechanical load and temperature variation respectively. $\boldsymbol{U}_{\mathrm{m}}$ and $\boldsymbol{U}_{\mathrm{T}}$ are the nodal displacement vector under only the mechanical load and thermal load, respectively. $\boldsymbol{K}$ is the structural stiffness matrix related to the design variable $\boldsymbol{x}$. Their detailed finite element formulations can be found in article [145].

Eqs. (5.23) and (5.24) are the equilibriums associated with mechanical and thermal loads. Eq. (5.25) is the constraint of the available base material volume fraction, where $\bar{V}$ and V are the available material volume fraction and the actually used material volume fraction respectively. ρ^{PAM} denotes the relative density of the porous anisotropic material, which is defined as

$$\rho^{\mathrm{PAM}} = \frac{\int_Y \boldsymbol{\rho} dy}{\mathrm{V}^{\mathrm{MI}}} \tag{5.29}$$

ρ^{PAM} can also be viewed as the volume fraction of material disposed on the micro scale. Together we define the volume fraction of material used on the macro scale ρ^{MA} as

$$\rho^{\mathrm{MA}} = \frac{\int_\Omega \boldsymbol{P} dy}{\mathrm{V}^{\mathrm{MA}}} \tag{5.30}$$

ρ^{MA} reflects the volume fraction of material disposed on the macro scale. And V^{MI} and V^{MA} are the volume of the micro design domain Y and macro design domain Ω respectively. Thus, the actually used material volume fraction V can be stated as

$$V = \rho^{\mathrm{PAM}} \cdot \rho^{\mathrm{MA}} \tag{5.31}$$

Note that here we do not specify the material volume in macrostructure and microstructure separately, which was done in previous study for static stiffness and dynamic response because the solid microstructure is not the optimal when thermo-load exists. Eqs. (5.26) and (5.27) sets bounds for density variables of two scales, where δ is a small predefined value to avoid singularity of the stiffness matrix. N and n are the total number of elements over the macro design domain Ω and micro design domain Y, respectively.

As we can see, the objective function is composed of two items. One represents the normalized structural compliance f_{c} when only the mechanical loads are applied on structures. While the other represents the normalized thermal expansion measure f_{u} of the area we have interests in when only the thermal loads are considered. In order to find the Pareto optimal of the bi-objective optimization problem, many approaches have been proposed in literatures. Among them, the most intuitive approach, though not the most rigorous, is the method of constructing a single aggregate objective function. The basic idea is to combine all the objective functions into a single functional form. A well-known combination is the weighted linear sum of the objectives. One specifies scalar weights for each objective to be optimized, and then combines them into a single function that can be solved by the single-objective optimizer. ω_c and ω_u in Eq. (5.22) are the scalar weights, which is determined by the experienced engineers. Moreover, the Pareto set can be obtained by varying the weight coefficients ω_{c} and ω_{u}. However, some potential issues may exist, such as the Pareto set can't be obtained via the approach above if the Pareto set is non-convex. Through further numerical tests, this problem does not present in the following numerical examples.

5.2.2 *Sensitivity analysis*

The sensitivity of the objective function with respect to the design variable x_i is comprised of two items

$$\frac{\partial f_{\mathrm{c}}}{\partial x_i} = -(\boldsymbol{U}_{\mathrm{m}})^{\mathrm{t}} \frac{\partial \boldsymbol{K}}{\partial x_i} \boldsymbol{U}_{\mathrm{m}} / C_{\mathrm{m}}^0 \tag{5.32}$$

and

$$\frac{\partial f_{\mathrm{u}}}{\partial x_i} = \frac{-2\boldsymbol{\Lambda}_\lambda^{\mathrm{t}} (\frac{\partial \boldsymbol{K}}{\partial x_i} \boldsymbol{U}_{\mathrm{T}} - \frac{\partial \boldsymbol{F}_{\mathrm{T}}}{\partial x_i})}{C_{\mathrm{T}}^0} \tag{5.33}$$

where $\boldsymbol{\Lambda}_\lambda$ is the adjoint response vector and is determined by the following equations (see Appendix E for detailed derivations).

$$\begin{gathered} \boldsymbol{K}\boldsymbol{\Lambda}_\lambda = \boldsymbol{\beta}_\lambda \\ \boldsymbol{\beta}_\lambda = \left\{0,0,\ldots,\boldsymbol{U}_{\mathrm{T}}^{\lambda(i)},0,0,\ldots 0\right\}^{\mathrm{t}} \end{gathered} \tag{5.34}$$

Since

$$\boldsymbol{K} = \sum_{e=1}^{\mathrm{N}} \int_{\Omega_e} \boldsymbol{B}_e^{\mathrm{t}} \boldsymbol{D}^{\mathrm{MA}} \boldsymbol{B}_e d\Omega \tag{5.35}$$

$$F_{\mathrm{T}} = \sum_{e=1}^{\mathrm{N}} \int_{\Omega_e} \boldsymbol{B}_e \boldsymbol{D}^{\mathrm{MA}} \boldsymbol{\kappa}^{\mathrm{MA}} \Delta T d\Omega$$

where $\boldsymbol{B}$ is the strain-displacement matrix, $\boldsymbol{D}^{\mathrm{MA}}$ and $\boldsymbol{\kappa}^{\mathrm{MA}}$ are the macro elastic constitutive matrix and the thermal expansion coefficient matrix of the porous material. It is worth mentioning that since $\boldsymbol{D}^{\mathrm{MA}} = P^{\alpha}\boldsymbol{D}^{\mathrm{H}}$, where $\boldsymbol{D}^{\mathrm{H}}$ represents the homogenized elastic constitutive matrix of the porous material, $\boldsymbol{\kappa}^{\mathrm{MA}}$ is not interpolated to avoid double penalization of the material, so we have $\boldsymbol{\kappa}^{\mathrm{MA}} = \boldsymbol{\kappa}^{\mathrm{H}}$. According to reference (Liu and Cheng [146]; Hashin [147]), no matter what configuration micro unit cell has, we always have $\boldsymbol{\kappa}^{\mathrm{H}} = \boldsymbol{\kappa}^{\mathrm{B}}$ for porous anisotropic materials composed by one base material, where $\boldsymbol{\kappa}^{\mathrm{B}}$ being the thermal expansion coefficients matrix of the base material. Therefore, $\boldsymbol{\kappa}^{\mathrm{MA}} = \boldsymbol{\kappa}^{\mathrm{H}} = \boldsymbol{\kappa}^{\mathrm{B}}$, and we will not distinguish them in this section.

Now, the derivatives of the global stiffness matrix $\boldsymbol{K}$ and load vector $\boldsymbol{F}_{\mathrm{T}}$ w.r.t. the design variables in Eq. (5.35) can be stated as

$$\frac{\partial \boldsymbol{K}}{\partial P_i} = \sum_{e=1}^{\mathrm{N}} \frac{\partial \boldsymbol{K}_e}{\partial P_i} = \frac{\alpha}{P_i}\boldsymbol{K}_i \tag{5.36}$$

$$\frac{\partial \boldsymbol{F}_{\mathrm{T}}}{\partial P_i} = \sum_{e=1}^{\mathrm{N}} \frac{\partial \boldsymbol{F}_{\mathrm{T}}}{\partial P_i} = \frac{\alpha}{P_i}(\boldsymbol{F}_{\mathrm{T}})_i \tag{5.37}$$

$$\frac{\partial \boldsymbol{K}}{\partial \rho_i} = \sum_{e=1}^{\mathrm{N}} \int_{\Omega_e} \boldsymbol{B}_e^{\mathrm{t}} \frac{\partial \boldsymbol{D}^{\mathrm{MA}}}{\partial \rho_i} \boldsymbol{B}_e d\Omega = \sum_{e=1}^{\mathrm{N}} P_e^{\alpha} \int_{\Omega_e} \boldsymbol{B}_e^{\mathrm{t}} \frac{\partial \boldsymbol{D}^{\mathrm{H}}}{\partial \rho_i} \boldsymbol{B}_e d\Omega \tag{5.38}$$

$$\begin{aligned} \frac{\partial \boldsymbol{F}_{\mathrm{T}}}{\partial \rho_i} &= \sum_{e=1}^{\mathrm{N}} \int_{\Omega_e} \boldsymbol{B}_e \frac{\partial \boldsymbol{D}^{\mathrm{MA}}}{\partial \rho_i} \boldsymbol{\kappa}^{\mathrm{MA}} \Delta T d\Omega \\ &= \sum_{e=1}^{\mathrm{N}} P_e^{\alpha} \int_{\Omega_e} \boldsymbol{B}_e \frac{\partial \boldsymbol{D}^{\mathrm{H}}}{\partial \rho_i} \boldsymbol{\kappa}^{\mathrm{MA}} \Delta T d\Omega \end{aligned} \tag{5.39}$$

where $\partial(\boldsymbol{D}^{\mathrm{H}})/\partial \rho_i$ can be retrieved from literature [2, 111] as

$$\frac{\partial (\boldsymbol{D}^{\mathrm{H}})}{\partial \rho_i} = \frac{1}{|Y|} \int_Y (\boldsymbol{I} - \boldsymbol{b}^{\mathrm{t}}\boldsymbol{\chi})^{\mathrm{t}} \frac{\partial \boldsymbol{D}^{\mathrm{MI}}}{\partial \rho_i} (\boldsymbol{I} - \boldsymbol{b}^{\mathrm{t}}\boldsymbol{\chi}) dy \tag{5.40}$$

where χ is generalized displacements and $\boldsymbol{I}$ is the identity matrix. Combining with the relation $\boldsymbol{D}^{\mathrm{MI}} = \rho^{\alpha}\boldsymbol{D}^{\mathrm{B}}$, the above equation leads to

$$\frac{\partial(\boldsymbol{D}^{\mathrm{H}})}{\partial\rho_i} = \alpha\rho_i^{\alpha-1}\int_{Y_i}(\boldsymbol{I}-\boldsymbol{b}_i^{\mathrm{t}}\chi_i)^{\mathrm{t}}\boldsymbol{D}^{\mathrm{B}}(\boldsymbol{I}-\boldsymbol{b}_i^{\mathrm{t}}\chi_i)dy \tag{5.41}$$

where $\boldsymbol{b}_i$ is the stain-displacement matrix of the ith element in unit cell design domain Y.

5.2.3 *Numerical treatments*

(1) Numerical procedure

In this research, the volume preserved Heaviside projective described in section of 4.1.2 and the GCMMA method (globally convergent version of the method of moving asymptotes) [148] are utilized as the optimizer. The major numerical procedures are briefly summarized as follows.

(a) Initialize macro and micro artificial densities (design variables).
(b) Using the linear density filter and the volume preserving Heaviside filter to compute the physical densities of each element with Eqs. (4.47)-(4.34).
(c) Solve the homogenization problem to get the effective properties of the porous material.
(d) Quantify the response of the structure under mechanical load in Eq. (5.23) and thermo load in Eq. (5.24).
(e) Solve the adjoint vector using Eq. (5.34).
(f) Compute the objective, constraints and their sensitivities with respect to the physical densities using Eqs. (5.22), (5.25), (5.28), (5.32) and (5.33).
(g) Compute the sensitivities of the objective and constraints with respect to the artificial densities using Eq. (4.41).
(h) Update design variables using GCMMA. If convergence, go to Step 9. Otherwise, go to Step 2.
(i) Output the final topology result. Stop.

The simulations have been done by the authors with extensions in OOFEM [149], which is open source finite element code with object oriented architecture for solving mechanical, transport and fluid mechanics problems. We add one optimization class containing the design variables, constraints, objective, and optimization solvers to solve the topology optimization problem. The analysis class and the optimization class contain each other's reference and can easily access to and communicate with each other. In order to improve the computation efficiency, the preconditioned conjugate gradient iterative solver instead of the direct solver is employed. The incomplete LU decomposition with no fill up preconditioner is used to reduce the condition number of the global stiffness matrix in order to accelerate the convergence rate. What's more, the compressed column storage format is used for the storage of the sparse global stiffness matrix.

5.2.4 *Numerical examples*

In this section, three thermoelastic examples are presented. In the first example, we consider a plane stress elliptical sandwich structure. We want to design the configuration of the macro sandwich core and the microstructure of the porous material based on our concurrent multi-objective optimization model. And the second example extends the first example to axisymmetric case, where we solve an axial symmetric problem on the macro scale and a 3D homogenization problem on the micro scale. The last example concerns with 3D structures both on the macro and micro scales.

The base material used in the three examples is the CSi ceramic material with properties: Young's modulus $\mathrm{E} = 450\ \mathrm{GPa}$, Poisson's ratio $\upsilon = 0.16$ and thermal expansion coefficient $\kappa = 4.2 \times 10^{-6}$.

(1) **Numerical example 1**

Fig. 5.10 shows the geometry and boundary conditions of an elliptical sandwich structure whose interior and exterior surfaces both with thickness $0.03\ \mathrm{m}$ are non-designable domain. The vertical mechanical pressure $\mathrm{P} = 50000\ \mathrm{N/m}$ applies on the upper surface of the structure, and the temperature variation ascends from $800\ ^{\circ}\mathrm{C}$ to $1600\ ^{\circ}\mathrm{C}$ in the positive Y-direction. The macro design domain and non-designable domain are meshed into 712 and 204 elements (8-node bilinear quadrilateral element) respectively, and the micro material unit cell is meshed into 625 (8-node bilinear quadrilateral element) elements.

Therefore, there are a total of 1337 design variables. The target of optimal design is to minimize the structural compliance under the mechanical load and the thermo deformation of the points located on the upper surface of the structure under thermo load. In the initial design, all the design variable values equal to 0.5, and $C_c^0 = 3.69$, $C_u^0 = 0.01218$.

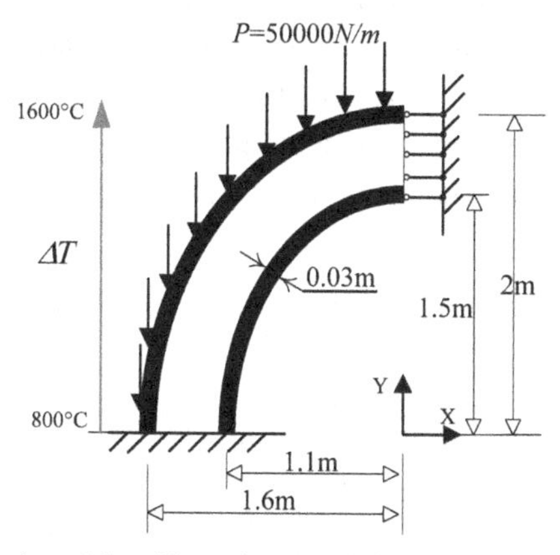

Fig. 5.10. Geometrical model and boundary conditions for numerical example 1 and 2

(a) Investigation of the influence of $\bar{V}$.

Table 5.4 shows the optimized designs of example 1 with the increase of $\bar{V}$ when $\omega_c = \omega_u = 0.5$.

1) Fig. 5.11 depicts the iteration history of objective value, constraint value V, micro material volume fraction ρ^{PAM} and macro material volume fraction ρ^{MA} for $\bar{V} = 0.3$. As can be observed in the optimization iteration process, ρ^{PAM} increases at the first few iterations and then decreases, while ρ^{MA} decreases and then increases, and the objective decrease with the iteration process when V reaches its upper bound 0.3 already. This reflects one of advantages of the proposed concurrent optimization model: the designs of macro and micro scales interact with each other, thus the distribution of materials on the two scales can be adjusted automatically to optimize the structural performances by the optimization model.

2) From Table 5.4 and also from Fig. 5.12, we can see that the normalized structural compliance f_c, the normalized thermal expansion measure of external surface f_u and hence the objective all decrease with

the increase of the available base material volume fraction $\bar{V}$ when $\bar{V}$ is less than 0.6. However, when $\bar{V}$ exceeds 0.6, the material volume fraction constraint Eq. (5.25) is no longer active. This demonstrates that $\bar{V}$ is not "the larger the better" for the bi-objective thermoelastic optimization problem. Therefore, there exists an "optimized" $\bar{V}$ which is about 0.6 in this numerical example. When $\bar{V}$ exceeds the "optimal" volume, the addition of material cannot improve the objective any more. This means that, when both structural resistance to mechanical load and its capability to resist the thermal expansion are emphasized, the lightweight design is not merely impelled by the economic factors but also by the requirement of improving the performance of structures.

Table 5.4. Investigation of the influence of $\bar{V}$ in example 1

$\bar{V}$	Volume fraction		Objective		Macro topology	Micro topology with 4×4 arrays
0.1	V	0.0997	*obj*	0.6525		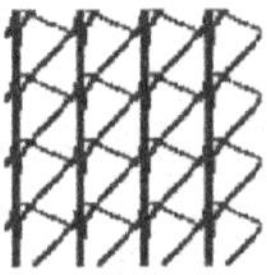
	ρ^{PAM}	0.2880	C_c	1.4054		
	ρ^{MA}	0.3463	C_u	0.01126		
0.3	V	0.2976	*obj*	0.4670		
	ρ^{PAM}	0.5290	C_c	0.3646		
	ρ^{MA}	0.5625	C_u	0.01024		
0.5	V	0.4978	*obj*	0.4404		
	ρ^{PAM}	1.000	C_c	0.1953		
	ρ^{MA}	0.4978	C_u	0.01008		
0.6	V	0.5142	*obj*	0.4392		
	ρ^{PAM}	1.000	C_c	0.1939		
	ρ^{MA}	0.5142	C_u	0.01006		
0.7 - 1.0	Identical to the case of $\bar{V}$=0.6 and the material volume fraction constraint is not active in these cases.					

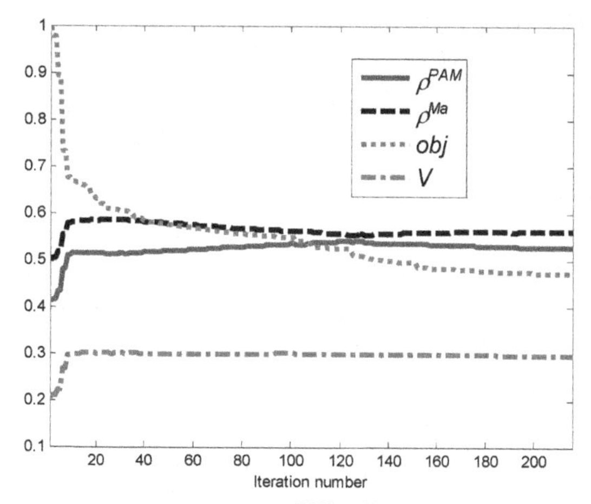

Fig. 5.11. Iteration history curves of ρ^{PAM}, ρ^{MA}, obj and V when $\bar{V} = 0.3$

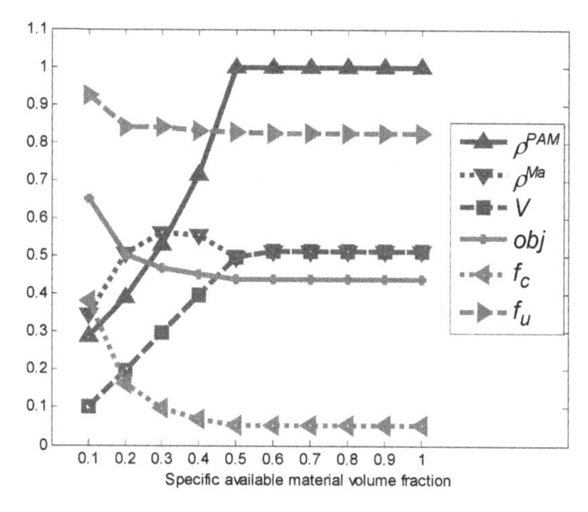

Fig. 5.12. Change of ρ^{PAM}, ρ^{MA}, V, obj, f_{c} and f_{u} with the increase of $\bar{V}$

This phenomenon can be explained by analyzing the sensitivity expressions. As can be seen from Eq. (5.32), the sensitivity of f_{c} is negative, which means f_{c} decreases with the increase of design variables. However, the sensitivity of f_{u} in Eq. (5.33) cannot always be guaranteed to be negative, which means f_{u} may increase with the increase of design variables. Therefore, the sensitivity of the total weighted objective is not always negative, and this may suggest the available material volume fraction is not the more the better. This can be verified by the setting $\bar{V} =$

1.0 and both the macro and micro design variables equal to 1, we obtain $C_c = 0.0597$, $C_u = 0.01275$ and $obj = 0.5319$. By comparing 0.5319 with its counterparts in Table 5.4 (0.6525 of $\bar{V} = 0.1$, 0.4647 of $\bar{V} = 0.3$, 0.4404 of $\bar{V} = 0.5$, 0.4392 of $\bar{V} = 0.6$), we can find it's not always wise to add material in the multi-objective thermoelastic design.

What's more, if we consider a two bar thermoelastic structure subject to uniform temperature increase ΔT and vertical mechanical force F illustrated in Fig. 5.13. A_1 is fixed and we find an optimum $A_2\,(0 \leq A_2 \leq 2A_1)$ to minimize objective. To minimize f_m, it should satisfy that $A_2 = 2A_1$. To minimize f_u, it should satisfy that $A_2 = A_1$. Therefore, to minimize obj $(\omega_c, \omega_u \in (0,1))$, A_2 should satisfy $0 \leq A_2 \leq 2A_1$. This means A_2 is not "the larger the better". See detail deduction in Appendix E.

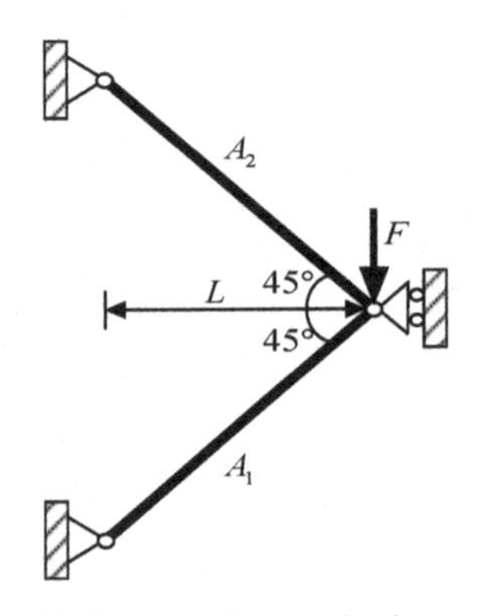

Fig. 5.13. Two bar thermoelastic structure

Table 5.5. Results of single macro scale multi-objective optimization of example 1

$\bar{V}$	0.1	0.3
obj	0.6610	0.4736
C_c	0.9973	0.2975
C_u	0.0128	0.0106
Structural Topology		

3) When $\bar{V}$ is less than 30%, the optimized configuration of material microstructure is porous, which means the porous material is advantageous to the solid material when the amount of the available material is not sufficiently given. And in order to demonstrate the merits of utilizing the porous material in thermoelastic problems, we solve the multi-objective optimization problem of example 1 in the single macro scale, where the material is specified as solid isotropic. The results are listed in Table 5.5. By comparing with the results in Table 5.4, we can see the structural compliance C_{c} under mechanical load in Table 5.5 is lower than the one in Table 5.4, but the thermo deformation measure of the structure C_{u} under thermal loads in Table 5.5 is higher than that in Table 5.4. This means porous material is conducive to reduce the thermo deformation caused by the thermo load. The mechanical load has influence on the configuration of the material microstructure since the optimized material microstructures in Table 5.4 have reinforced skeletons in Y-direction to resist the uniform pressure in Y-direction. When $\bar{V}$ exceeds 50%, the optimized micro material configuration is isotropic solid, which indicates the utilization of the solid material to construct the macro structure is instrumental to decrease the total objective. Our concurrent optimization model can give satisfying designs on structural and material scales covering isotropic and anisotropic materials.

(b) Investigation of the influence of ω_{c} and ω_{u}

Table 5.6 shows the optimized designs of example 1 with different weighting coefficients ω_{c} and ω_{u} when $\bar{V} = 0.5$.

a) When $\omega_{\mathrm{c}}:\omega_{\mathrm{u}} = 1:0$, branch-like structure appears in the macro designable domain and more material is disposed in the upper macro designable domain to resist the mechanical force. The optimized micro material configuration is isotropic solid, which means isotropic solid material is advantageous to porous material in the sense of minimizing the compliance of the structure subject to only mechanical load. When $\omega_{\mathrm{c}}:\omega_{\mathrm{u}} = 0.5:0.5$, more material is used to connect the interior and exterior surface sheets and coordinate their deformation caused by thermo load in order to minimize f_{u}. When $\omega_{\mathrm{c}}:\omega_{\mathrm{u}} = 0:1$, the optimized micro material configuration is porous. With the increase of ω_{u}, C_{u} becomes smaller and C_{c} becomes larger. By varying ω_{c} and ω_{u}, experienced engineers can design a structure according to the relative importance of the influences of mechanical loads and thermal loads on the structure.

Table 5.6. Investigation of the influence of ω_{c} and ω_{u} when $\bar{V} = 0.5$ in example 1

$\omega_c : \omega_u$	Objective		Structural topology	Micro Topology 4×4 arrays
1:0	obj	0.4821		
	C_{c}	0.1101		
	C_{u}	0.01138		
0.99:0.01	obj	0.4686		
	C_{c}	0.1170		
	C_{u}	0.01103		
0.5:0.5	obj	0.4404		
	C_{c}	0.1953		
	C_{u}	0.01008		
0:1	obj	1.2599		
	C_{c}	6.7034		
	C_{u}	0.00859		

2) The macro structure topology of $\omega_{\mathrm{c}}:\omega_{\mathrm{u}} = 0.99:0.01$ shows obvious difference with that of $\omega_{\mathrm{c}}:\omega_{\mathrm{u}} = 1:0$. Here the phenomenon may reflect that the optimal design of the thermoelastic problem is sensitive to the temperature variation. In some cases, the optimal structural topology changes drastically when a small temperature variation is introduced besides the mechanical load. This phenomenon is also observed in articles [54, 150] where structural stiffness and thermal conductivity are considered as objectives. The optimal design can be dominated by the temperature variation which is very small, especially for stiff structures.

This suggest that using single criteria, for example the compliance, to account both influences of the mechanical load and the thermo load on the thermoelastic structures may not be appropriate, because the influence of the thermo load on the structure may far exceeds that of the mechanical load. Therefore, it is a better way to consider the influence of the mechanical load and the influence of the thermo load on the structure separately. What's more, the temperature environment where most engineering structures serve in always changes from time to time; it's difficult for us to design a structure that is optimal for different temperature variations. In our multi-objective model, we consider the influences of the thermo load and the mechanical load on the structure separately. And f_{c} and f_{u} are both normalized, therefore the optimization model is less sensitive to the magnitudes of the mechanical load or thermal load.

(2) **Numerical example 2**

This example extends example 1 to axisymmetric case. The geometry and boundary conditions are exactly the same as those in example 1. But the macro structure is axisymmetric, and the model is described in cylindrical coordinates (X,θ, Y) in this example. So, in Fig. 5.12, X-axis now represents the coordinate in the radius direction, and Y-axis represents the coordinate in the axial symmetric direction. We solve an axial symmetric problem on the macro scale and a 3D homogenization problem on the material scale. The mesh of the macro structure is the same as that in numerical example 1. However, we need to be more careful because there are many differences here, for example, the volume of the macro finite element should consider its distance from the symmetric axis and the elastic constitutive matrix should be extended to axial symmetric form. The micro design domain is a cube unit cell, which is meshed into $15 \times 15 \times 15$ linear 3D eight-node finite elements. Therefore, there are totally 4087 design variables. In the initial design, all the design variable values equal to 0.5, and $C_{\mathrm{c}}^{0} = 2.7 \times 10^{-1}$, $C_{\mathrm{u}}^{0} = 7.6 \times 10^{-3}$. $\omega_{\mathrm{c}} = \omega_{\mathrm{u}} = 0.5$. The optimized results with the increase of $\bar{V}$ are summarized in Table 5.7.

By comparing the macro structures in Table 5.7 with those in Table 5.4, we can find more material is distributed around the symmetric axis when the macro structure is axisymmetric, because there is vertical pressure on the top of the outer surface, and the macro structure needs to be reinforced there. As to the micro structure, when $\bar{V} = 0.3$, more material is distributed to reinforce the micro structure in the radius direction X and the circumferential direction θ. What's more, the optimal configuration of material microstructure is porous, which means the utilization of porous material is advantageous to the solid material, when the light design of structures is emphasized.

Table 5.7. Results of example 2

$\bar{V}$	Objective		Structural Topology	Micro topology : one cell and 2×2×2 arrays
0.3	obj	0.5826		
	C_c	0.0830		
	C_u	0.0066		
0.5	obj	0.5339		
	C_c	0.0630		
	C_u	0.0064		

(3) **Numerical example 3**

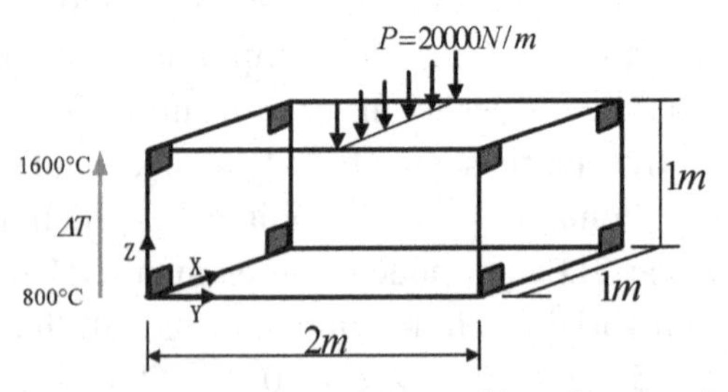

Fig. 5.14. Geometrical model and boundary conditions for numerical example 3

As shown in Fig. 5.14, the $2\ \text{m} \times 1\ \text{m} \times 1\ \text{m}$ block is fixed at two side face of $\text{Y} = 0,2$. The vertical mechanical pressure $\text{P} = 20000\ \text{N/m}$ in the negative Z-direction applies on the line $(\text{Y}, \text{Z}) = (1,1)$. And the temperature variation ascends from $800°C$ to $1600°C$ in the positive Z-direction. Due to the two symmetry planes, only a quarter of the structure is analyzed and optimized, i.e., only the domain of coordinates ($X \leq 0.5$, $Y \leq 1.0$, Z) with the plane of $X = 0.5$ fixed in the X-direction and the plane of $Y = 1.0$ fixed in the Y-direction. The macro 1/4 model considered is meshed equally into 12, 20 and 12 finite element divisions in the X-, Y- and Z-directions, respectively. And the finite elements attached to surfaces $Z = 0, 1$ compose the non-designable domain. Therefore, there are totally 2400 design variables on the macro scale. The micro design domain is a cube cell which is meshed into $15 \times 15 \times 15$ linear 3D eight-node finite elements. Therefore, there are totally 5775 design variables. The objective of optimal design is to minimize the structural compliance under the mechanical load and the thermo deformation of the points located on the surface $Z = 1$ under the thermo load. In the initial design, all the design variable values equal to 0.5, and $C_c^0 = 32.95$, $C_u^0 = 1.35 \times 10^{-3}$. $\omega_c = \omega_u = 0.5$. The optimized results of the two scales with the increase of $\bar{V}$ are summarized in Table 5.8.

Table 5.8. Results of example 3

$\bar{V}$	Objective		Macro topology	Micro topology: one cell and 2×2×2 arrays
0.3	obj	0.4749		
	C_c	0.2870		
	C_u	0.6630		
0.5	obj	0.3563		
	C_c	0.1050		
	C_u	0.6080		

From Table 5.8, we can see that the structural compliance C_c, the thermal expansion measure of external surface C_u and hence the objective obj all decrease with the increase of the available base material volume fraction $\bar{V}$. And the configurations of the macro and micro structures will change obviously with the increasing of $\bar{V}$. More material is distributed in the Z-direction to resist the vertical pressure observed from the macro scale, and the reinforced pillars of microstructure transforms from oblique to vertical with the increase of $\bar{V}$ observed from the micro scale. What's more, the optimal configuration of material microstructure is also porous in this example, which means the utilization of porous material is helpful to increase the capacity to resist the thermal expansion of the load carrying structures.

5.3 *Concluding remarks*

In this chapter, multiscale concurrent topology optimization of material and structural design under mechanical and thermal loads is considered. Corresponding problem formulation and numerical solution procedures are also developed for finding the optimal solutions. It is found that compared with the case where only mechanical load is involved, anisotropic porous material with well-designed microstructures are very effective to reduce the structural compliance when both mechanical and thermal loads are applied. Analytical benchmark example validates the results obtained through numerical procedure.

In addition, using a single criterion, for example the compliance, to account for both the influences of the mechanical load and the thermo load on the thermoelastic structures may not be appropriate. Hence, the multi-objective optimization model, which considers the influences of the mechanical load and the influences of the thermo load on the thermoelastic structures separately, is recommended. Optimal porous micro material configurations are observed in some multi-objective optimization examples, especially when the available amount of material is insufficiently given (the structural weight is specified strictly). This indicates that, sometimes the utilization of porous materials in the multi-objective optimization model is more advantageous than that of the solid isotropic material. When both the structural compliance and thermo deformation are considered in the objective function, an "optimal" material volume fraction is observed in some cases. When the available material volume fraction reaches this “optimal” material volume fraction, we cannot minimize the objective further by increasing the amount of available material. The extraordinary properties of porous material present in thermoelastic design exhibit a promising applications for multi-disciplinary field, which deserves further studies.

In this work, only global structural response (i.e., compliance and thermal deformation of concerned domain) is accounted for. However, some local behaviors may exit, e.g., the stress concentration at the concave corner, and the significant thermal deformation of a small region in local. It is also very important to improve the local mechanical behaviors above as well some others such as thermal stress reduction and thermal buckling load enhancement through multiscale design and optimization. These works are currently under investigation and we will report the results in the future works.

Chapter 6

Optimization of Fiber-Reinforced Composite Materials and Structures

This chapter discusses the problem of concurrent multi-scale design optimization of the material and structure of fiber-reinforced composites (i.e., composite plate, shell, and frames) by considering the characteristics of the discrete fiber ply and winding angle due to the manufacturing requirements. For composite plate and shell structure, to gain the best match of the macroscale topology and microscale material selection, the fiber ply angle in one element can be considered as one microscale design variable, and the materials that exist or not in the design domain are considered as the macroscale design variable in the content of multiscale optimization of composite structures. For composite frame structure, the fiber winding angle in the micro-material scale and the geometrical parameter of components in the macro structural scale are introduced as the independent variables to realize multiscale design optimization of composite frames.

To overcome the non-differentiable objective function caused by discrete ply angles, a continuously treatment based on the Discrete Material Optimization (DMO) method is introduced. However, the traditional DMO method easily results in a not clear fiber angle selection, thus the Heaviside projection is introduced and the new HPDMO (Heaviside Penalization Discrete Material Optimization) method is then proposed. Meanwhile, the measure of convergence ratio with respect to ply angles is introduced. The numerical examples of GFRP/CFPR lamina show that the proposed method can significantly enhance the convergence ratio and the distinct ply angle selections are thereby obtained. Based on that, under the framework of the PAMP scheme, the concurrent multiscale optimization approaches with respect to the composite frames are proposed to improve the structural stiffness and dynamic frequency. Numerical examples show that the proposed frameworks can well consider the trade-off between the coupled micro and macro scales, and optimal ply angles at the microscale and distribution of structured material composed with glass or carbon fiber reinforced polymers (GFRP/CFRP) can be obtained, simultaneously.

By the way, in the practical engineering, a ply angle sequence that meet the manufacturing constraints can significantly eliminate the risk of structural failures. So, six kinds of constraints, i.e., contiguity constraint, 10% rule, balance constraint, damage tolerance constraint, symmetry constraint, DMO normalization constraint are explicitly included in the optimization model as a series of linear inequalities or equalities. The capabilities of the proposed optimization model are demonstrated with the example of compliance minimization and fundamental frequency maximization in following sections, subject to a constraint on the composite volume. There's no doubt that the concurrent multiscale frame work of fiber-reinforced composite presented in this chapter can be extended to explore stress constraint, frequency constraint, and multi-physics problems.

The work of this chapter is mainly related with references [151-154].

6.1 *The HPDMO method and convergence ratio*

6.1.1 *Discrete material interpolation of ply angle*

The variable stiffness optimization of laminates or composite frame structures composed of fiber reinforced polymer (FRP) will be studied in the following. According to the conceptions in the previous chapter, the optimization problem of laminate has the nature of multiscale characteristics, see Fig. 6.1 for details. At the macroscale, the optimal distribution of structured material composed of FRP will be explored. At the microscale, the optimal selection of FRP ply angles will be determined. Similar to the traditional integer programming problem, the ply angle is selected from a discrete set of ply angles to ensure the manufacturing requirements [152, 155]. Generally, candidate ply angles set may be $[0^\circ, \mp 45^\circ, 90^\circ]$. Here, the non-gradient-based or gradient-based algorithms can be employed to solve this micro optimization problem. It is common that the gradient-based algorithms are helpful to enhance the optimization efficiency, and thereby will be employed below. However, the discrete ply angle will result in a non-differentiable objective function. Therefore, a continuously treatment is needed. In the present study, the DMO method is introduced.

The DMO method [72, 97] can be regarded as a generalized multi-material penalty/interpolation model, which selects one material from a given set of candidate materials for one element in the macro-structure to satisfy the manufacturing constraint and optimize the objective function. Thus, the ply angle of the composite materials can be optimized. It should be noted that the void material is introduced into the set of candidate materials in this study to realize the change of the macro structural topology. Then, considering the coupling effect, the concurrent optimization of the composite materials and structures can be realized for the specified structural loading and boundary conditions. The problem is encompassed in the same parameterization and solved in the finite-element framework.

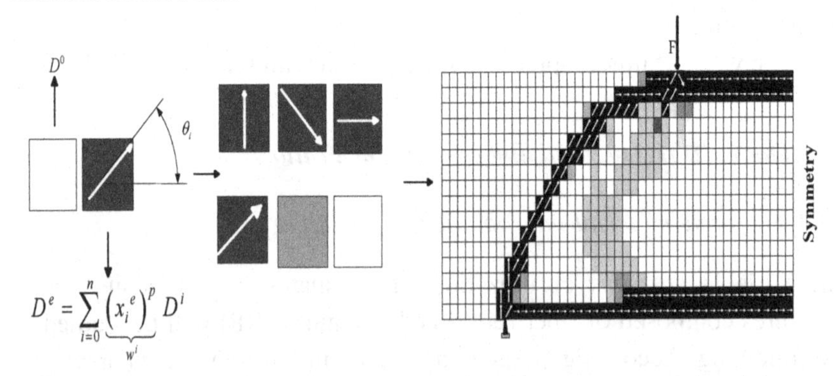

Fig. 6.1. Illustration of the concurrent optimization of composite materials and structures

In DMO the elemental constitutive matrix $\boldsymbol{D}_i^{\mathrm{e}}$ (the superscript e refers to "element in macroscale") can be expressed as a weighted sum of the candidate materials, which is characterized by a constitutive matrix $\boldsymbol{D}_{i,j}$ as

$$\begin{aligned}\boldsymbol{D}_i^{\mathrm{e}} &= \sum_{j=1}^{\mathrm{n}} w_{i,j}\boldsymbol{D}_{i,j} \\ &= w_{i,1}\boldsymbol{D}_{i,1} + w_{i,2}\boldsymbol{D}_{i,2} + \cdots + w_{i,n}\boldsymbol{D}_{i,\mathrm{n}} \\ &\quad 0 \le w_{i,j} \le 1 \quad i \in \mathrm{N}, \ j \in \mathrm{n}\end{aligned} \tag{6.1}$$

where n is the number of candidate materials in one element, and N is the number of elements in the macro structural domain. The total number of design variables for a single layered structure is $\mathrm{N} \times \mathrm{n}$. $w_{i,j}$ in Eq. (6.1) is the weight of the *j*-th material of the *i*-th element, which takes a value between 0 and 1. This weight coefficient is defined as

$$\begin{gathered} w_{i,j} = \left(x_{i,j}\right)^p \prod_{k=1}^{\mathrm{n}} \left[1 - \left(x_{i,k\neq j}\right)^p\right] \\ 0 \le x_{i,j} \le 1, j \in \mathrm{n} \qquad \sum_{j=1}^{n} x_{i,j} = 1 \quad i \in \mathrm{N} \end{gathered} \tag{6.2}$$

where p is the penalty index, $\boldsymbol{D}_i^e$ is the elastic constitutive matrix of the *i*-th element, $\boldsymbol{D}_{i,j}$ is the elastic constitutive matrix of the *j*-th candidate material of the ith element, and $x_{i,j}$ is the design variable of the *j*-th candidate material of the *i*-th element. If $x_{i,j} = 1$, the *i*-th element chooses the *j*-th material from the set of candidate materials. If $x_{i,j} = 0$, the *i*-th element does not contain the *j*-th candidate material. It should be noted that, if $x_{i,j}$ takes the median value of 0 and 1, the sum of $x_{i,j}$ for *i*-th element is not equal to 1 in Eq. (6.1), then the normalized weight strategy as Eq. (6.3) should be adapted. To push the weights towards 0 and 1 and avoid a fuzzy ply angle selection, Lund and Stegmann [85] used the following interpolation of the elemental constitutive matrix $\boldsymbol{D}_i^{\mathrm{e}}$ from $\boldsymbol{D}_{i,j}$,

$$\boldsymbol{D}_i^{\mathrm{e}} = \sum_{j=1}^{\mathrm{n}} \overline{w}_{i,j}\boldsymbol{D}_{i,j}, \tag{6.3}$$

where

$$0 \le x_{i,j} \le 1 \quad i \in \mathrm{N}, \qquad j \in \mathrm{n}$$

$$\overline{w}_{i,j} = \frac{w_{i,j}}{\sum_{j=1}^{n} w_{i,j}} = \left[\frac{1}{\sum_{j=1}^{n} w_{i,j}} \left(x_{i,j}\right)^{p} \prod_{k=1}^{n}\left[1 - \left(x_{i,k\neq j}\right)^{p}\right]\right], \quad (6.4)$$

Note that

$$\sum_{j=1}^{n} \overline{w}_{i,j} = 1. \quad (6.5)$$

6.1.2 *The HPDMO model and numerical treatments*

(1) Continuous penalty strategy

The strategy of continuous [156, 157] index p in Eq. (6.3) is adopted in this section. The initial penalty index p is given as $p = 1$. When the iteration of optimization satisfies a certain convergence criterion, it gradually increases until the optimization results converge. The model is labeled as CPDMO (Continuous Penalization Discrete Material Optimization) in this section. The numerical examples show that the CPDMO model can help to reduce the probability that the optimization process converges very fast to the local optimal solution compared with the traditional DMO model, and effectively improve the convergence rate of the optimization results.

(2) Improved DMO model based on Heaviside projection (HPDMO)

Unlike the traditional DMO model in Eq. (6.3), we introduce the modified Heaviside penalty function shown as Eq. (6.6) into the DMO material interpolation formula to help the design to make discrete choices. For the detailed descriptions of the Heaviside penalty function, please refer to the reference of Guest et al. [102], where the Heaviside filter was originally used as a density filter in topology optimization. Sigmund [104] proposed a modified Heaviside filter. Xu et al. [130] presented a volume-preserving Heaviside filter by combining the Heaviside filter and the modified Heaviside filter and adding a parameter in the new filter for controlling the volume change by the filter. Guest et al. [157] investigated the effect of constant penalty β in the Heaviside filter on the resulting design and convergence history.

Taking advantage of Heaviside projection, the design variable is interpolated as

$$\overline{x} = e^{-\beta(1-x)} - (1-x)e^{-\beta}, \tag{6.6}$$

and the weight coefficient is defined as

$$w_{i,j} = \bar{x}_{i,j}\prod_{k=1}^{\mathrm{n}}\left(1 - \bar{x}_{i,k\neq j}\right). \tag{6.7}$$

Then a normalization is introduced and it yields

$$\overline{w}_{i,j} = \frac{w_{i,j}}{\sum_{j=1}^{\mathrm{n}} w_{i,j}}, \tag{6.8}$$

Finally, we have

$$\boldsymbol{D}_i^{\mathrm{e}} = \sum_{j=1}^{n} \overline{w}_{i,j}\boldsymbol{D}_{i,j} \tag{6.9}$$

where $x_{i,j}$ is the artificial density of candidate materials, $\overline{x_{i,j}}$ is the density after the nonlinear penalty, and β is the parameter of the nonlinear penalty.

Fig. 6.2 shows the curve of the nonlinear function Eq. (6.6) with different penalty parameters β. As observed from this figure, as β increases, the penalty of the nonlinear function significantly increases, which drives the density after penalty to 0 or 1 and ultimately prevents the design from falling into the difficulties of cross-fiber selection. The cross-fiber considered in this section is defined as the case when there are two or more candidate materials in one element. The normalization of weights ensures that the sum of $\overline{w}_{i,k}$ keeps 1 and only one candidate material is finally selected as β increases. However, the existence of cross-fiber introduces difficulties to the manufacture and application of the optimal result. Consequently, one of the key factors of the discrete optimization of composite material is how to avoid the cross-fiber difficulty and obtain a clear optimal result. Note that although increasing the value of β can help to punish the design variables of either 0 or 1. However, excessive penalty of the design variables will apparently make the optimization iterations controlled by the nonlinear interpolation function, and the optimization process will not effectively reflect the requirements of the structural performances. So a continuous penalty strategy [156, 157] is suggested to be adopted.

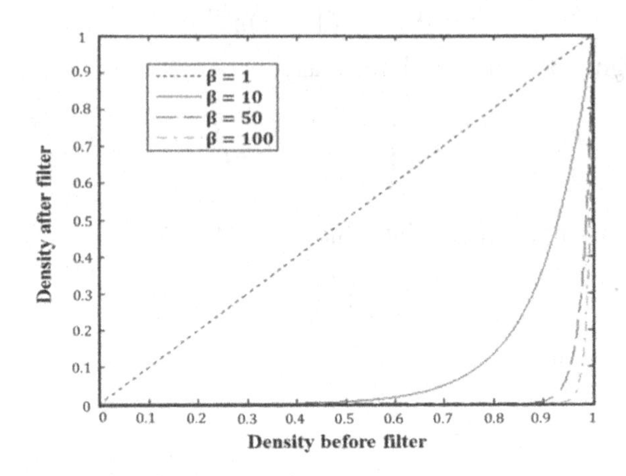

Fig. 6.2. Penalty of the Heaviside function with different β values

(3) Convergence assessment criterion

To determine whether the optimization has converged to a satisfactory result, i.e., a single candidate material has been chosen in a specified element, and all other materials have been discarded. The convergence criterion is defined as Eq. (6.10) [85]. For each element, the following inequality is evaluated according to all weight factors of ith element as.

$$w_j \geq \varepsilon\sqrt{w_1^2 + w_2^2 + \cdots + w_n^2} \tag{6.10}$$

where ε is a tolerance level; typically, $\varepsilon \in [0.95\sim0.99]$. If inequality Eq. (6.10) is satisfied for any w_j in the element, the element is flagged as converged. The convergence assessment criterion is measured by the convergence ratio H_ε, and it is defined as the ratio of converged elements for a total number of elements, i.e., $H_\varepsilon = N_{\text{tol}}/\text{N}$, where N_{tol} is the number of convergent elements. If the tolerance level ε is 95%, and the optimization fully converges, i.e., $H_{\varepsilon=0.95} = 1$, all the elements have a single weight that contributes more than 95% to the Euclidian norm of the weight factors.

(4) Suggestions of parameter selection in HPDMO

To ensure a stable iteration, a continuous penalty strategy [156, 157] with a small initial value of β is used, which prevents the optimization process from falling into the local optimal solution too early. Simultaneously, a continuous penalty strategy is also employed and the penalty parameter β is increased by a small value in each update. According to the convergence assessment criterion in Eq. (6.10), the code can determine whether the element has been converged and determine the convergence rate [102] of the entire structure.

For the practical application, the optimization model can judge the absolute value of the convergence rate of H_ε before and after the optimization. Note that the slope value of β is problem-dependent: in our examples, we suggest linearly increasing the penalty index with a slope of 1.01, and in 1.2 to 1.5 at the initial stage of optimization iteration and when the convergence rate difference is less than the convergence index, respectively, according the numerical experiences of the examples in this chapter. In our examples, according to the numerical experiences, we suggest adopting linear mechanism of increasing notably low slope at the beginning and determining whether to use a larger slope to gradually increase the penalty index to 200 according to the convergence criteria.

6.2 *Structural analysis and maximizing stiffness optimization of the laminate*

6.2.1 *Analysis of composite structures characterized by discrete materials*

The finite element method is used to obtain the response of the composite plane and shell structures that are subjected to a given set of loading and boundary conditions. The elemental stiffness matrix of element i, $\boldsymbol{K}_i^{\mathrm{e}}$, is obtained as the integration of the constitutive matrix of candidate materials in the element, as shown in Eq. (6.11).

$$\boldsymbol{K}_i^{\mathrm{e}} = \int_{\Omega^i} \boldsymbol{B}^{\mathrm{T}} \boldsymbol{D}_i^{\mathrm{e}} \boldsymbol{B} d\Omega^i \tag{6.11}$$

The element stiffness matrix is used to assamble the structural stiffness matrix in the linear static equilibrium $\boldsymbol{KU} = \boldsymbol{F}$, where $\boldsymbol{U}$ and $\boldsymbol{F}$ are nodal vectors of the displacements and external forces, respectively. Hence, the compliance C, which is the objective function of the optimization can be stated as $\mathrm{C} = \boldsymbol{U}^{\mathrm{T}}\boldsymbol{F} = \boldsymbol{U}^{\mathrm{T}}\boldsymbol{KU}$.

6.2.2 *Optimization model based on the discrete material constraint*

The concurrent optimization of the material and structure of composites based on the minimum structural compliance can be expressed as the following mathematical model.

$$\text{find} \quad \boldsymbol{X} = \{x_{i,j}\} \tag{6.12}$$

$$\min \quad C = \boldsymbol{F}^{\mathrm{T}}\boldsymbol{U} = \boldsymbol{U}^{\mathrm{T}}\boldsymbol{KU} \tag{6.13}$$

$$\text{s.t.} \quad \boldsymbol{KU} = \boldsymbol{F}$$

$$V(x_{i,j}) = \sum_{i}^{\mathrm{N}} \sum_{j}^{\mathrm{n-1}} w_{i,j}(x_{i,j}) v_i \leq f_{\mathrm{v}} \overline{V} \tag{6.14}$$

$$0 \leq x_{\min} \leq x_{i,j} \leq 1$$

where v_i is the macro volume of the ith element, and the void material doesn't count in the volume constraint. $\overline{V}$ is the entire volume of the macro-design domain, f_{v} is the volume fraction of fibers, and $x_{\min}$ ($x_{\min} = 0.001$ in this chapter) is a small positive value to avoid singularity of global stiffness matrix during the optimization iterations. Furthermore, the void material is introduced into the set of candidate materials to effectively realize the optimal distribution of fiber material within the macro structural design domain and to realize the topological optimization of the macro-structure.

Sensitivity analysis

Sensitivity analysis is important for the gradient-based optimization algorithm to improve computational efficiency. The elemental stiffness matrix can be expressed as shown in Eq. (6.11). From $\boldsymbol{KU} = \boldsymbol{F}$, the sensitivity of the structural compliance and the volume constraint with respect to the design variables can be written as:

$$\frac{\partial C}{\partial x_{k,l}} = \sum_{i=1}^{\mathrm{N}} \sum_{j=1}^{\mathrm{n}} -\boldsymbol{U}_i^{\mathrm{T}} \frac{\partial \boldsymbol{K}_i^{\mathrm{e}}(x_{i,j})}{\partial x_{k,l}} \boldsymbol{U}_i \tag{6.15}$$

$$\frac{\partial \boldsymbol{V}(x_{i,j})}{\partial x_{k,l}} = \sum_{i=1}^{\mathrm{N}} \sum_{j=1}^{\mathrm{n}} \frac{\partial w(x_{i,j})}{\partial x_{k,l}} v_i \tag{6.16}$$

where, the corresponding derivatives of $\partial \boldsymbol{K}_i^{\mathrm{e}}/\partial x_{k,l}$ and $\partial w(x_{i,j})/\partial x_{k,l}$ are nonzero only when $i = k,\ j = l$. Because the strain–displacement matrix $\boldsymbol{B}$ is independent of the design variables, the sensitivities of the structural compliance and the volume constraint are only relevant to the material interpolation formula of $\boldsymbol{D}_i^e$ and the weight function of $w_{i,j}$. Then the derivative of the material interpolation formula of $\boldsymbol{D}_i^e$ can be expressed as follows.

$$\frac{\partial \boldsymbol{D}_i^{\mathrm{e}}}{\partial x_{i,l}} = \sum_{j=1}^{\mathrm{n}} \frac{\partial \overline{w}_{i,j}}{\partial x_{i,l}} \boldsymbol{D}_{i,j} \tag{6.17}$$

where the term $\partial \overline{w}_{i,j}/\partial x_{i,l}$ can be obtained from Eqs. (6.6)-(6.8).

The explicit sensitivity expressions of the objective function and the volume constraint on the design variables can be obtained by substituting $\partial \boldsymbol{D}_i^{\mathrm{e}}/\partial x_{i,l}$ and $\partial \overline{w}_{i,j}/\partial x_{i,l}$ into Eq. (6.15) and Eq. (6.16), respectively. In the study, the SQP method is used as the optimizer. The major numerical procedures are briefly summarized as follows.

(a) Initialize the candidate design variables in the structural elements $x_{i,j}$.
(b) According to the proposed material interpolation of Eq. (6.9), calculate the normalized weight function of $\overline{w}_{i,j}$ of each candidate material.
(c) Obtain the structural displacements under specified loads and boundary condition by solving $\boldsymbol{KU} = \boldsymbol{F}$.
(d) According to the displacement responses, calculate the objective function and the constraint functions using Eqs. (6.13)-(6.14), respectively.

(e) Using Eqs. (6.15)-(6.16), calculate the sensitivity of the objective function and the constraint function with respect to the design variables of the artificial density of candidate materials.
(f) Judge the absolute value of the convergence rate H_ε before and after the optimization. If the absolute value of the difference is less than a constant convergence index (suggested to be 1%), then we can increase the penalty index β.
(g) Solve the approximate optimization problem using the SQP algorithm and update the design variables of $x_{i,j}$. If converge, go to step (h); otherwise, go to step (b).
(h) Output the final concurrent optimization results. Stop.

6.2.3 *Optimization examples and discussions*

Based on the wide use of plane and shell structure in engineering, the following three examples are solved and the differences among the traditional DMO model, the CPDMO model, and the HPDMO model on the optimum value of the objective function, convergence rate, and iterative history are discussed. All examples allow for 5 candidate materials in each element, which contains 4 fiber candidate materials and 1 void material. The fiber candidate materials are glass-fiber-reinforced epoxy with the orthotropic properties $\mathrm{E_x} = 39\ \mathrm{GPa}$, $\mathrm{E_y} = 8.4\ \mathrm{GPa}$, $\upsilon_{\mathrm{xy}} = 0.26$, and $\mathrm{G_{xy}} = 4.2\ \mathrm{GPa}$. For easy manufacture, the four fiber candidate materials adopt the most commonly used fiber ply angle as $[0^\circ, \mp 45^\circ, 90^\circ]$. The initial values of the design variables are $x_{i,j} = 0.2, i \in \mathrm{N}, j \in \mathrm{n}$ i.e., the initial design of the candidate materials is uniformly distributed in the design domain. The tolerance level of the examples is $\varepsilon = 0.95$, and the convergence assessment criterion is $H_{\varepsilon=0.95}$. We can realize the macroscopic structural topology optimization by choosing the void material or specified fiber material for one element and simultaneously realize the micro optimization of the fiber ply angle by selection from the 4 candidate materials.

(1) **Example 1:** Concurrent multiscale optimization of MBB beam

An MBB beam with a concentrated force at the top edge is

investigated as a standard academic testing example for minimum compliance optimization with discrete material candidates. The structure and its geometric sizes are shown in Fig. 6.3. The design domain is divided into 20×80 meshes with eight-node planar elements. The volume constraint for the following three examples is set as $f_v = 65\%$.

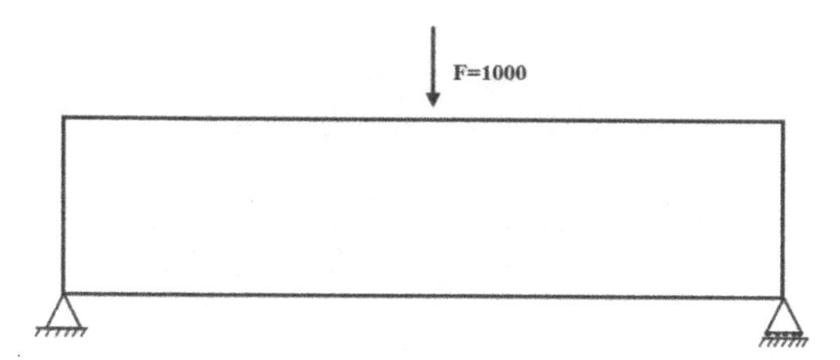

Fig. 6.3. Geometrical model and boundary conditions of an MBB beam

Fig. 6.4-Fig. 6.6 show the optimization results of the MBB beam with the concurrent optimization of materials and structures with three types of discrete material optimization model (DMO, CPDMO, and HPDMO). The Fig. 6.4-Fig. 6.6(a) are the macro topology results. The black areas represent the distribution of reinforced fiber materials, i.e., one type of the 4 fiber materials. The white area is the void material, and the gray areas are the elements with the cross-fiber. The Fig. 6.4-Fig. 6.6(b) show the microscopic material distribution results i.e., the optimal fiber ply angle. The straight lines in the elements indicate the fiber ply direction, and the white area is void material or cross-fiber element, which can be distinguished by comparing with its counterpart in the macro-topology figure. Table 6.1 shows the comparisons of the compliance and the convergence rate of the three discrete material optimization models.

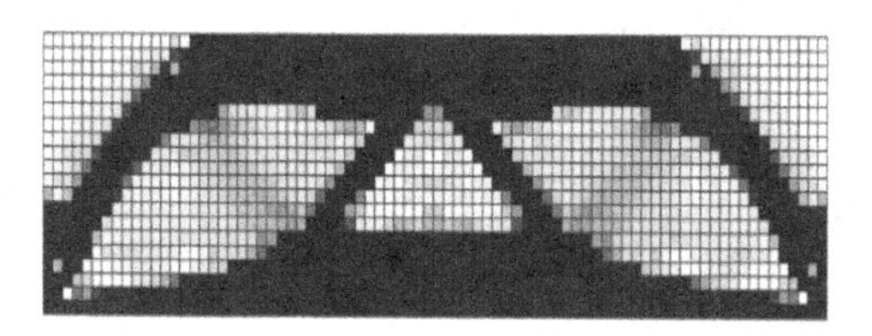

(a) Macro topology optimization results

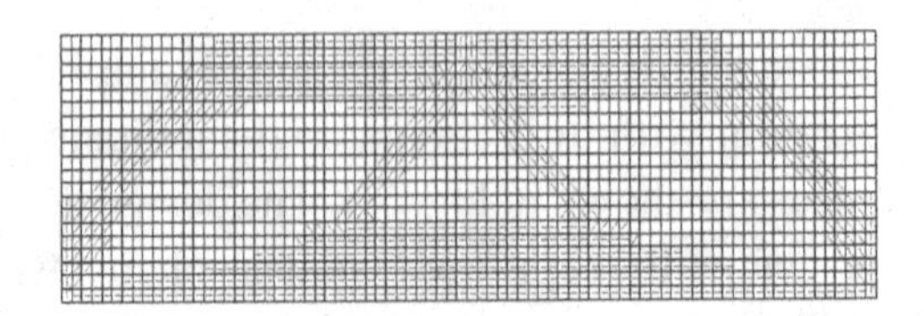

(b) Microscopic optimization result of the fiber ply angle

Fig. 6.4 Concurrent optimization design of MBB beam based on the DMO model

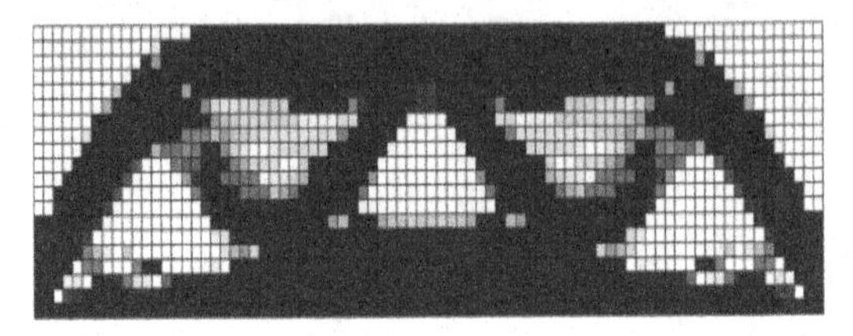

(a) Macro topology optimization results

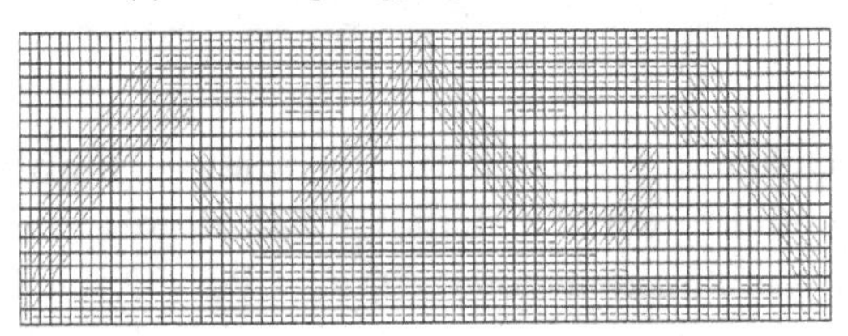

(b) Microscopic optimization results of the fiber ply angle

Fig. 6.5. Concurrent optimization design of the MBB beam based on the CPDMO model

(a) Macro topology optimization results

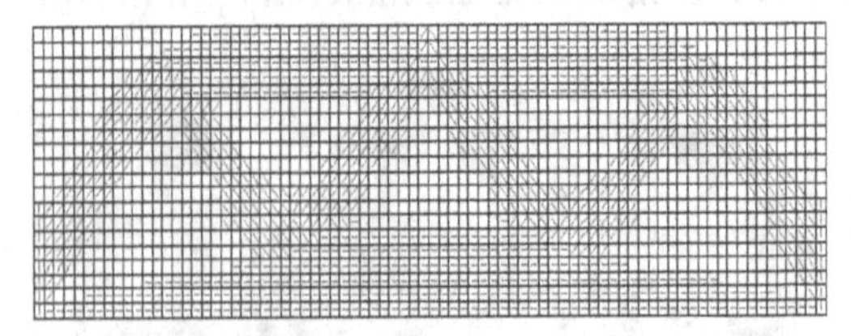

(b) Microscopic fiber ply angle optimization results

Fig. 6.6. Concurrent optimization design of the MBB beam based on the HPDMO model

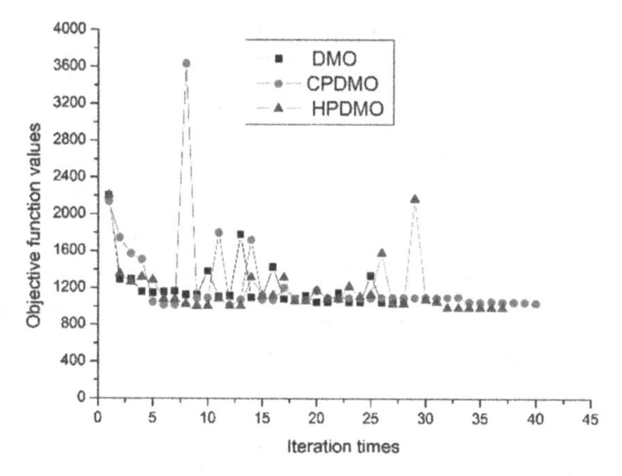

Fig. 6.7. Iteration history of the objective function

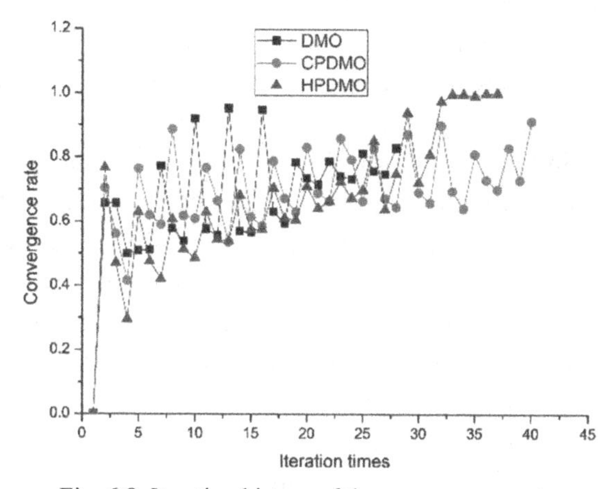

Fig. 6.8. Iteration history of the convergence rate

Table 6.1. The comparison of the optimization results with different penalty models of the MBB beam

Method	Compliance	Convergence rate $H_{\varepsilon=0.95}$
DMO	1046.4	83.0%
CPDMO	1034.2	91.2%
HPDMO	981.3	100.0%

(a) Based on the macro topology optimization results in Fig. 6.4-Fig. 6.6, the macroscopic structure configurations of the three types of material penalty models are notably similar to the result that is obtained with classical topology optimization technique [141]. However, the DMO and CPDMO models suffer from the "gray-scale elements", particularly in parts of the internal oblique supports. This problem indicates the difficulty of cross-fiber selection, which implies that the convergence rate is low in these elements and cannot provide clear fiber choices. Therefore, it will undoubtedly increase the difficulty and the cost of manufacture of the optimized design. However, Fig. 6.6 shows that the HPDMO model can provide a much clearer optimization result with the convergence rate (100%), which is favorable to manufacture the optimized results. Simultaneously, we observe from Table 6.1 that compared with the DMO and CPDMO models, the compliance of the HPDMO model decreases by 6.2% and 5.1% with the identical volume fraction, respectively.

(b) From the results of microscopic fiber ply angle optimization in Fig. 6.4-Fig. 6.6, the elements near to the upper and lower boundary of the design domain are composite with ply fiber in horizontal direction, because the principal stress of the elements is along the horizontal direction for this loading case. Although the parts of the internal oblique support, the DMO, and the CPDMO models cannot give a clear choice of the fiber ply angle, the improved HPDMO model gives the distribution of positive or negative 45 degrees of reinforced fibers to bear the shear stress, which is consistent with the loading characteristics of the MBB beam. To some extents, the results also validate the optimization models and algorithms of the HPDMO model.

(c) It should be noted that at the initial stage of the application of the HPDMO model, the optimization parameters value β in Eqs. (6.6)-(6.9) is set as a small value, and the penalty of the interpolation function is weak. This weak punishment can ensure the approximation of the interpolated material constitutive approaches to the physical one. With the progress of the iterations, the weight function of the candidate materials gradually approaches 0 or 1. Subsequently as the value of β increases pretty large, it finally drives the weight function to 0 or 1 to realize a clear choice of the fiber ply angle.

(d) From the optimization iteration history of Fig. 6.7 and Fig. 6.8, compared with the DMO model and the CPDMO model, the HPDMO model takes fewer iterations to achieve the desired optimization results.

(2) **Example 2:** Concurrent multiscale optimization of L-shaped beam

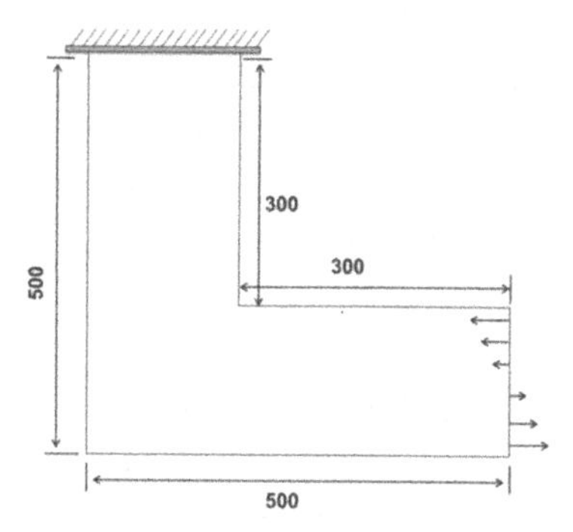

Fig. 6.9. Geometrical model and boundary conditions of the L beam

An L-shaped beam with relatively complex geometries and boundary conditions is investigated in this example. The loading/boundary conditions and the geometric size are shown in Fig. 6.9. The upper left boundary of the L-shaped beam is fixed, and the right end of the beam is applied with a bending moment. According to the Saint-Venant principle, the bending moment equivalently transfers to a linear distribution force along the border edge. The design domain is divided into 1600 elements with eight-node planar elements. The volume fraction of fibers is $f_{\mathrm{v}} = 56\%$. Fig. 6.10-Fig. 6.12 show the optimization results of the L-shaped beam with concurrent optimization of the composite material and structure based on the three types of material penalty model. The labeling denotes the same signification as of Example 1. Table 6.2 shows the comparison of the compliance and the convergence rate of the three types of material penalty model.

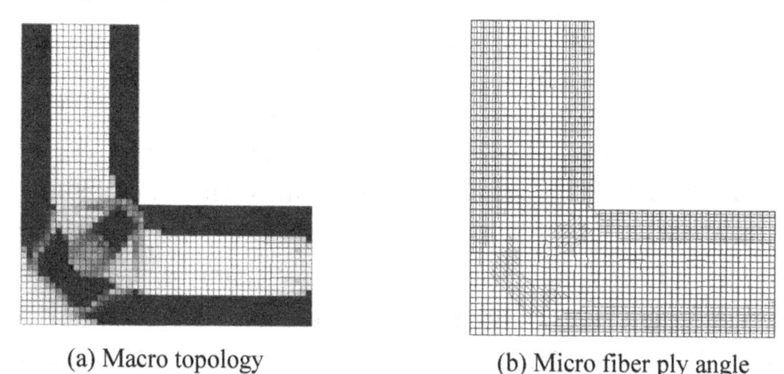

(a) Macro topology (b) Micro fiber ply angle

Fig. 6.10. Concurrent optimization design of the L beam based on the DMO model

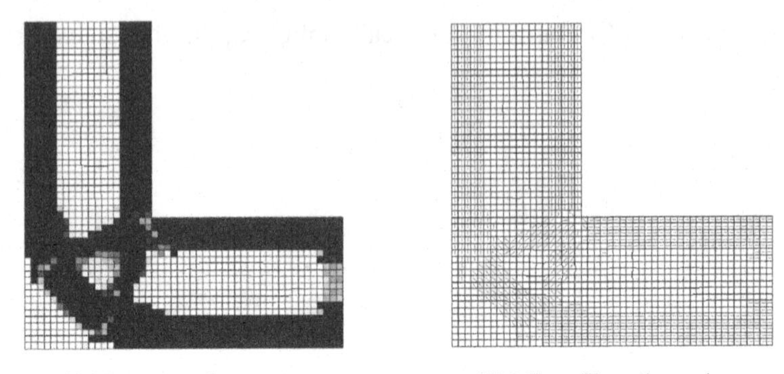

(a) Macro topology (b) Micro fiber ply angle

Fig. 6.11. Concurrent optimization design of the L beam based on the CPDMO model

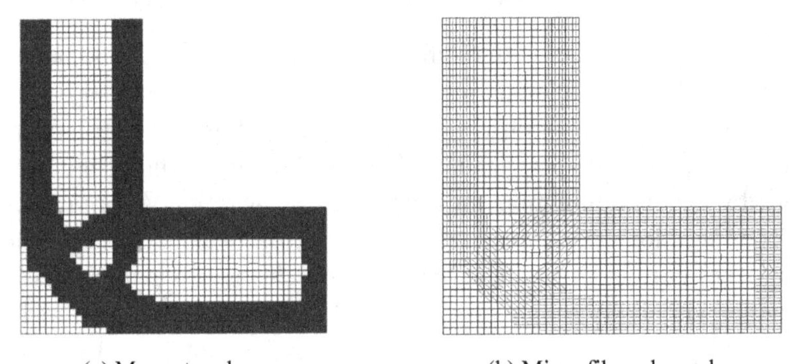

(a) Macro topology (b) Micro fiber ply angle

Fig. 6.12. Concurrent optimization design of L beam based on the HPDMO model

Table 6.2. The comparison of the optimization results with different penalty models of the L beam

Method	Compliance	Convergence rate ($H_{\varepsilon=0.95}$)
DMO	90427.8	90.50%
CPDMO	89344.6	97.88%
HPDMO	80703.0	100.0%

Fig. 6.10-Fig. 6.12(a) show that the optimized macrostructures based on the three material penalty models have almost identical configurations. The reinforced fiber materials (black units) are mainly distributed in the L beam left and right sides of the vertical section, the upper and lower sides of the horizontal section, and the overlapping areas of the horizontal and vertical sections. These distributions are consistent with that of the main

stress that corresponds to the bending moment of the L beams. Furthermore, the optimization results of the DMO model and the CPDMO model show the "gray-scale element" at the right end of the beam and the lower left overlapping section, which indicates that the material choice in these regions is not clear. However, the optimized results of the HPDMO model are relatively clear, which implies the $H_{\varepsilon=0.95}$ complete convergence.

From the micro fiber orientation based on the subgraphs (b) of Fig. 6.10-Fig. 6.12, for the three penalty models in the horizontal and vertical regions, the fiber ply angles are horizontal and vertical, respectively. Conformed to the mechanical analysis of the L-shape beam under the bending moment, the horizontal and vertical parts of the L-shape beam are subjected to the stress of tension and the pressure, respectively. The fiber ply angle can be effectively optimized using the HPDMO model in the previously mentioned area with complicated stress state, and the proposed HPDMO model can be adapted to a relatively complex structure such as the L-shaped beam.

The optimization iteration history of Fig. 6.13-Fig. 6.14 and the data in Table 6.2 show that the HPDMO model gives the lowest compliance value in the same volume fraction, whose objective function values decrease by 10.75%, and 9.67% compared that of the DMO and CPDMO models, respectively.

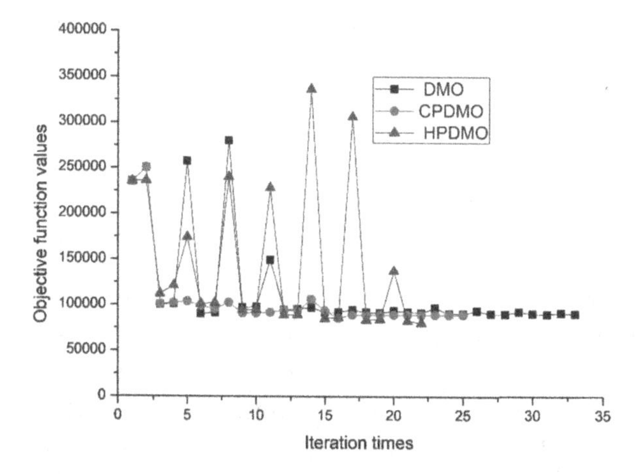

Fig. 6.13. Iteration history of the objective function

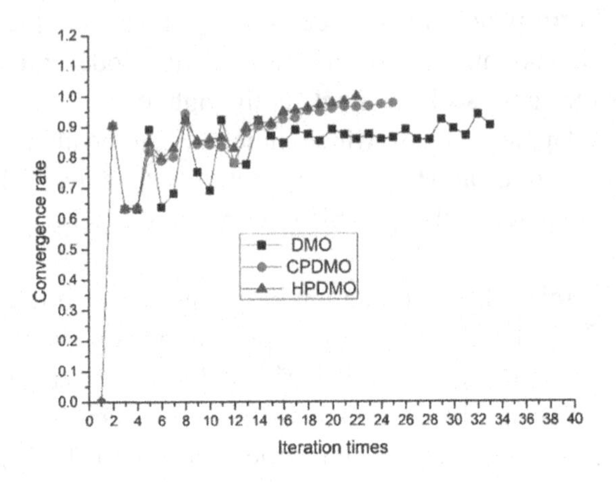

Fig. 6.14. Iteration history of the convergence rate

(3) **Example 3**: Concurrent multiscale optimization of a composite cylindrical shell

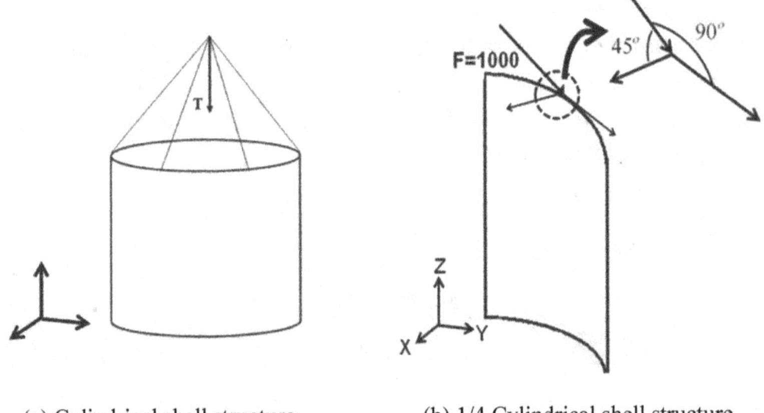

(a) Cylindrical shell structure (b) 1/4 Cylindrical shell structure

Fig. 6.15. Cylindrical shell and simplified 1/4 model

The third example aims to optimize a composite cylindrical shell, whose structure is shown in Fig. 6.15 with a radius of 0.5m and a height of 1m. The bottom of the cylindrical shell is fixed, and the top of geometric center is applied with a concentrated force through the transmission of 4 rigid rods. Because of rotational symmetry of the structural geometry and the load with rotating 90 degree, the problem is simplified to a quarter model of the cylindrical shell. Examples 1 and 2 discussed the pros and cons of the DMO, CPDMO and HPDMO models; therefore, this example only considers the improved HPDMO model to perform the concurrent optimization design of a spatial composite shell. The volume constraint is set as $f_V = 66\%$.

To conveniently check the optimization results, Fig. 6.16 shows the ichnography of the optimization results of the macro and material distribution topology. The labels denote the same significances as those of Example 1.

From the macro optimized structural topology and the fiber ply angle shown in Fig. 6.16, we conclude that the optimization results also have 90° rotational symmetry, which satisfies the geometric structure and the premise of the load application to some extent. Similar to the optimization results of the HPDMO model in Examples 1 and 2, the optimization results in this example also have a clear macro structural topology and choices of microscopic fiber ply angle.

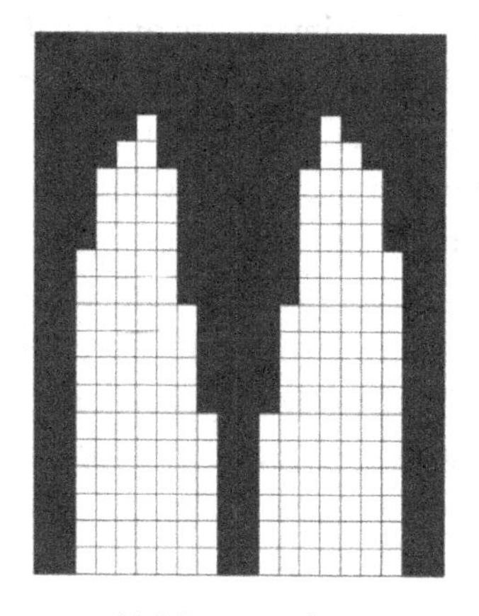

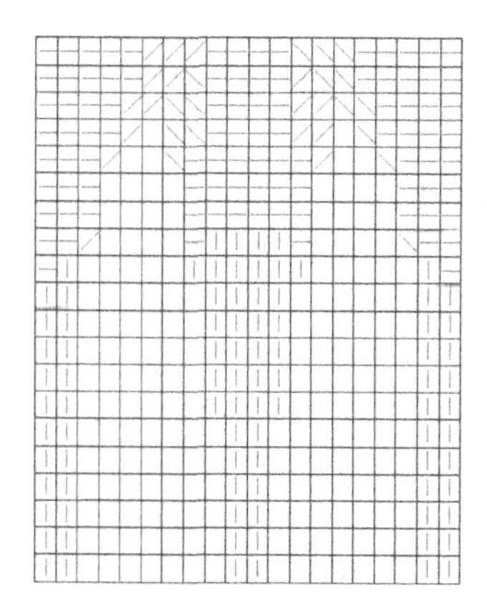

(a) Macro topology (b) Micro fiber ply angle

Fig. 6.16. Concurrent optimization design of a cylindrical shell structure (1/4 model) based on the HPDMO model

In Fig. 6.15 the fastigiated load can be decomposed into a horizontal force and a vertical force, so the entire cylindrical shell sustains a strong downward pressure and surface tension. Therefore, the fiber distribution in the vertical direction satisfies the demands of transferring loads, and the horizontal and vertical fibers can resist the surface tension well.

It is worth noting that, the example of a cylindrical shell with concurrent optimization of material and structure can be easily extended to apply in composite grid structures. In the engineering tendons, a grid structure is visualized as the main load-bearing structure, and the skin that wraps outside the tendons plays a weak role in the load-carrying structure. Therefore, the orientation and arrangement of the tendons definitely affect the mechanical properties of the grid structure. The proposed discrete material interpolation scheme of the HPDMO model in this section can clearly simulate the ply angles and ply position of the tendons, which provides a reference for the innovative structural design of a new composite grid structure.

6.3 *Concurrent multiscale stiffness optimization of composite frame structures*

6.3.1 *Concurrent multiscale optimization of composite frame structures*

In most cases, the structural configuration optimization and material design are carried out independently, which generally requires many iterations, costing a lot of time and computational resources. For a composite frame, the micro fiber ply parameters (i.e., fiber winding angle, layer thickness, and ply stacking sequence) and the macro structural characteristics (i.e., cross-sectional area of components and topology configuration) can definitively affect the stiffness performance of the structure. So, in this section, a concurrent multiscale optimization model is established, which can fully consider the coupling effect of macro and micro scale design variables and take advantage of the potential of composite structures.

(1) Conception of concurrent multiscale optimization of composite frames

The illustration of concurrent multiscale optimization of the composite frame structure is shown in Fig. 6.17. For easy of deduction and no loss of generality, we assume the composite frame structure studied in this section is composed with circular tubes. At the macroscale, the radius of cross-section is recognized as macro design variable. As in classical topology optimization of the frame structure, the tube's radius can be recognized as the size and topology variables at the same time. When the radius reaches its lower bound, the tube can be regarded to be deleted from the ground structure to realize the structural topology optimization if the lower bound is small enough. It should be pointed out specially that to avoid the heavy burden of reconstruction of the stiffness matrix, the tube with very little radius is kept in the finite element model in numerical implementation. So, it can be restored with the optimization iterations, but in the current step the tube has little contribution to the stiffness of the frame structure since its radius is already approaching its lower bound, a very little value. At the microscale, the fiber winding angle (i.e., $\theta_{i,j}$ for continue optimization model and $x_{i,j,k}$ for discrete optimization model) is recognized as the micro design variable, and in the present research a uniform fiber winding angle is adapted in the same layer. Considering the constraint of the practical manufacturing, the fiber winding angle is assumed to be the same in the same layer. A detailed discussion about the continuous or discrete micro-material model for the design variable of microscale fiber winding angle has been given in Section 6.1.

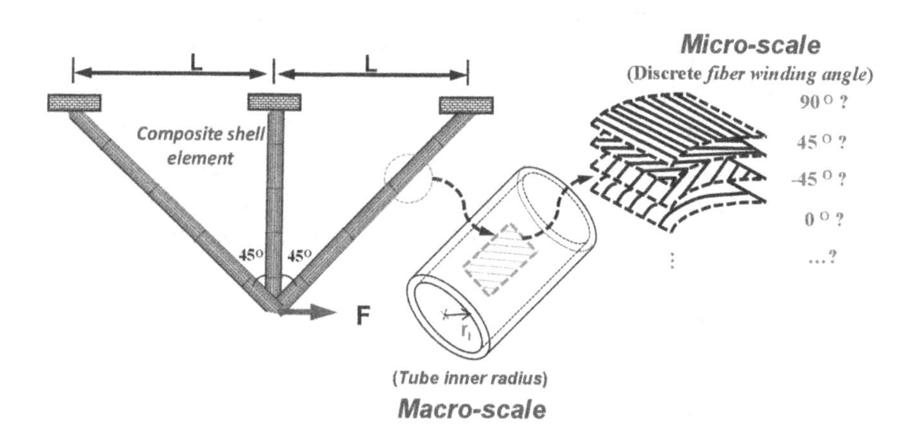

Fig. 6.17. Illustration of the concurrent optimization of a composite frame

In Sections 6.3-6.4, the classical composite laminate theory, which will be presented in Section 6.3.3, is adopted to achieve the structural and sensitivity analysis of composite frames. In Section 6.5, the response of the composite frames is analyzed based on an extension of the beam finite element tool called BEam Cross Section Analysis Software (BECAS) developed by Blasques and Lazarov [158] for anisotropic and inhomogeneous beam sections with arbitrary geometry.

(2) Optimization formulation with consideration of a discrete fiber winding angle

Considering the constraint of manufacturing, the fiber winding angle used in practical engineering is generally with discrete character, which means the candidate fiber winding angles are restricted to a pre-specified set of candidate fiber winding angles, such as $[0^\circ, \mp 45^\circ, 90^\circ]$. It will bring difficulty for the solution of the optimization of composite frame structure with gradient based algorithms if we introduce the winding angle as the design variables directly. So, some special treatment must be taken to cope with the difficulty in the optimization model. Based on the minimum structural compliance, the concurrent multiscale optimization of the composites frame in this case can be expressed as Eqs. (6.18)-(6.20). The structural compliance C is the objective function, which can be stated as $C = \boldsymbol{U}^{\mathrm{T}}\boldsymbol{K}\boldsymbol{U}$, and the structural volume is recognized as material constraint, where $\boldsymbol{U}$ and $\boldsymbol{K}$ are global vectors of the displacements and global stiffness matrix, respectively. The micro-scale design variable is set as the artificial candidate material density of $x_{i,j,k}$, where the subscript i, j and k means the number of the tube, layer, and the number of candidate materials, respectively. For the convenience of description, we label this Discrete fiber Concurrent Multiscale Optimization model as DCMO.

$$\text{find} \quad \boldsymbol{X} = \{x_{i,j,k}, r_i\} \;\; i \in \mathrm{N}^{\mathrm{tub}}, j \in \mathrm{N}^{\mathrm{lay}}, k \in \mathrm{N}^{\mathrm{can}} \tag{6.18}$$

$$\min \quad C = \boldsymbol{U}^{\mathrm{T}}\boldsymbol{K}\boldsymbol{U} \tag{6.19}$$

$$\begin{aligned} \text{s.t.} \quad & \boldsymbol{K}(\boldsymbol{D}(\boldsymbol{X}))\boldsymbol{U} = \boldsymbol{F} \\ & V(r_i) = \sum_{i=1}^{\mathrm{N}^{\mathrm{tub}}} \pi \left[{t_{\mathrm{tot}}^{i}}^{2} + 2 r_i t_{\mathrm{tot}}^{i} \right] \mathrm{L}_i \le f_{\mathrm{v}} \bar{V} \\ & 0 \le x_{i,j,k} \le 1 \\ & 0 \le r_{\min} \le r_i \le 0.1 \end{aligned} \tag{6.20}$$

In Eq. (6.18), $\mathrm{N}^{\mathrm{tub}}$ is the number of tubes in the frame structure. $\mathrm{N}^{\mathrm{can}}$ is the number of candidate materials in each layer. $\mathrm{N}^{\mathrm{lay}}$ is the number of layers in each tube, and in this section the number of layers i.e., $\mathrm{N}^{\mathrm{lay}}$ is assumed as the same in all tubes. The macro-scale design variable r_i is the inner radius of the tube. L_i is the length of the tube, t_{tot}^i is the total thickness of the tube. In the present section, we only consider the macro radius and micro fiber winding angle as the design variables, so the fiber layer thickness is constant in the optimization process. $\bar{V}$ is the entire volume of the macro-design domain, f_{v} is the volume fraction of materials. r_{min} is the lower limit of the tube's radius and generally adopts a small positive value. In the presented section, $r_{\mathrm{min}} = 0.1\mathrm{mm}$. We consider four kinds of fiber winding angles $\left[0^{\circ}, \mp 45^{\circ}, 90^{\circ}\right]$ (cf. [159]) as the candidate materials, and adopt HPDMO interpolation scheme [101, 160] to solve the discrete micro-scale optimization problem. In Eq. (6.18), if $x_{i,j,k} = 1$, that indicates the *i*-th tube's *j*-th layer chooses the *k*-th material from the set of candidate materials; and if $x_{i,j,k} = 0$, the *i*-th tube's *j*-th layer does not contain the *k*-th candidate material. More details about HPDMO formulation can be found in Section 6.1.2.

With a few differences from the DCMO model, if the fiber winding angle ($\theta_{i,j}$) can change continually within the range of $[-90°, 90°]$ and be directly recognized as the design variable, then the design variable $x_{i,j,k}$ in Eq. (6.18) of the DCMO model can be changed to $\theta_{i,j}$ in the mathematical formulation of the Continuous fiber Concurrent Multiscale Optimization model (label as CCMO).

6.3.2 *Parameterization model for discrete composite material optimization*

In this section, the fiber winding angle is assumed as the same in the same layer. So, the total number of design variables for a composite frame structure in DCMO model is $\mathrm{N}^{\mathrm{tub}} \times \mathrm{N}^{\mathrm{lay}} \times \mathrm{N}^{\mathrm{can}}$. Stegmann and Lund [161] extended the interpolation scheme to multi-materials form, which considered the coupling effect between the artificial densities. For composite frame structure, the interpolation model can be expressed as follows Eqs. (6.21)-(6.23) according to the discussion in Section 6.1.

$$\bar{x}_{i,j,k} = e^{-\beta(1-x_{i,j,k})} - (1 - x_{i,j,k})e^{-\beta} \tag{6.21}$$

$$w_{i,j,k} = \bar{x}_{i,j,k} \prod_{m=1}^{\mathrm{N^{can}}} (1 - \bar{x}_{i,j,m\neq k}) \tag{6.22}$$

$$\boldsymbol{D}^{i,j} = \sum_{k=1}^{\mathrm{N^{can}}} \underbrace{\left(\frac{w_{i,j,k}}{\sum_{p=1}^{\mathrm{N^{can}}} w_{i,j,p}} \right)}_{\overline{w}_{i,j,k}} \boldsymbol{D}_{i,j,k} \tag{6.23}$$

where $\bar{x}_{i,j,k}$ is the artificial density after the nonlinear penalty, and β is the parameter of the nonlinear penalty. Moreover, the convergence criterion is defined as Eq. (6.24), similarly. For each layer, the following inequality is evaluated according to all weight factors.

$$\overline{w}_{i,j,k} \geq \varepsilon \sqrt{\overline{w}_{i,j,1}^2 + \overline{w}_{i,j,2}^2 + \cdots + \overline{w}_{i,j,\mathrm{N^{can}}}^2} \tag{6.24}$$

where ε is a tolerance level; typically and $\xi \in [0.95\sim0.99]$ is suggested If inequality in Eq. (6.24) is satisfied for any $\overline{w}_{i,j,k}$ in the layer, the layer is flagged as converged. The convergence assessment criterion H_ε is defined as the ratio of converged layers for a total number of layers: $H_\varepsilon = N_{\mathrm{c}}^{\mathrm{l,tol}}/\mathrm{N^{l,tol}}$, where $N_{\mathrm{c}}^{\mathrm{l,tol}}$ is the total number of convergent layers, $\mathrm{N^{l,tol}}$ is the total number of layers.

6.3.3 *Structural analysis of composite frames and sensitivity analysis*

In this section, the composite frame is modeled using shell element and the classical composite laminate theory is used to model the laminate.

(1) Laminate stiffness with shell elements

According to the classical composite laminate theory, the stress in the *j*-th layer can be expressed in terms of the laminate middle-surface strains and curvatures as in Eq. (6.25).

$$\begin{bmatrix} \sigma_x \\ \sigma_y \\ \sigma_{xy} \end{bmatrix}_j = \begin{bmatrix} \bar{Q}_{11} & \bar{Q}_{12} & \bar{Q}_{16} \\ \bar{Q}_{12} & \bar{Q}_{22} & \bar{Q}_{26} \\ \bar{Q}_{16} & \bar{Q}_{26} & \bar{Q}_{66} \end{bmatrix}_j \left\{ \begin{bmatrix} \varepsilon_x^0 \\ \varepsilon_y^0 \\ \gamma_{xy}^0 \end{bmatrix} + z \begin{bmatrix} K_x \\ K_y \\ K_{xy} \end{bmatrix} \right\} \tag{6.25}$$

where ε_x^o, ε_y^o, γ_{xy}^o are the middle-surface strains, K_x, K_y are the bending curvature of the middle-surface, K_{xy} is the twist curvature of the middle-surface. z is the coordinate of the j-th layer along thickness. The transformed reduced stiffness $\overline{\mathrm{Q}}_{pq}, p, q \in 1,2,6$ can be given in the term of the reduced stiffness of Q_{pq} as defined in Eqs. (6.28)-(6.29). For convenience, we define laminated invariant parameters $\Gamma_1 - \Gamma_4$, which can be expressed as Eqs. (6.26)-(6.27).

$$\Gamma_1 = \frac{(3\mathrm{Q}_{11} + 3\mathrm{Q}_{22} + 2\mathrm{Q}_{21} + 4\mathrm{Q}_{66})}{8}$$
$$\Gamma_2 = \frac{(\mathrm{Q}_{11} - \mathrm{Q}_{22})}{2} \tag{6.26}$$

$$\Gamma_3 = \frac{(\mathrm{Q}_{11} + \mathrm{Q}_{22} - 2\mathrm{Q}_{12} - 4\mathrm{Q}_{66})}{8}$$
$$\Gamma_4 = \frac{(\mathrm{Q}_{11} + \mathrm{Q}_{22} + 6\mathrm{Q}_{12} - 4\mathrm{Q}_{66})}{8} \tag{6.27}$$

For the orthotropic material, the reduced stiffness Q_{11}, Q_{22}, Q_{12}, Q_{66} can be expressed as the function of engineering constants as Eqs. (6.28)-(6.29).

$$\mathrm{Q}_{11} = \frac{\mathrm{E}_{11}}{1 - \mu_{12}\mu_{21}}, \mathrm{Q}_{22} = \frac{\mathrm{E}_{22}}{1 - \mu_{12}\mu_{21}} \tag{6.28}$$

$$\mathrm{Q}_{12} = \frac{\mu_{21}\mathrm{E}_{11}}{1 - \mu_{12}\mu_{21}} = \frac{\mu_{12}\mathrm{E}_{11}}{1 - \mu_{12}\mu_{21}}, \mathrm{Q}_{66} = \mathrm{G}_{12} = \mathrm{G}_{31} \tag{6.29}$$

where, E is the Young's modulus, G is the shear modulus, μ is the Poission's ratio. Then the transformed reduced stiffness $\overline{\mathrm{Q}}_{pq}$ can easily be expressed as Eqs. (6.30)-(6.31):

$$\begin{Bmatrix} \bar{Q}_{11} \\ \bar{Q}_{22} \\ \bar{Q}_{12} \\ \bar{Q}_{66} \\ \bar{Q}_{16} \\ \bar{Q}_{26} \end{Bmatrix}^j = \begin{bmatrix} \Gamma_1 & \Gamma_2 & 0 & \Gamma_3 & 0 \\ \Gamma_1 & -\Gamma_2 & 0 & \Gamma_3 & 0 \\ \Gamma_4 & \Gamma_2 & 0 & -\Gamma_3 & 0 \\ \frac{1}{2}(\Gamma_1 - \Gamma_4) & 0 & \frac{1}{2}\Gamma_2 & 0 & \Gamma_3 \\ 0 & 0 & \frac{1}{2}\Gamma_2 & 0 & \Gamma_3 \\ 0 & 0 & \frac{1}{2}\Gamma_2 & 0 & -\Gamma_3 \end{bmatrix} \begin{Bmatrix} 1 \\ \cos(2\theta_{i,j}) \\ \sin(2\theta_{i,j}) \\ \cos(4\theta_{i,j}) \\ \sin(4\theta_{i,j}) \end{Bmatrix} \tag{6.30}$$

$$\Gamma = \begin{bmatrix} \Gamma_1 & \Gamma_2 & 0 & \Gamma_3 & 0 \\ \Gamma_1 & -\Gamma_2 & 0 & \Gamma_3 & 0 \\ \Gamma_4 & \Gamma_2 & 0 & -\Gamma_3 & 0 \\ \frac{1}{2}(\Gamma_1 - \Gamma_4) & 0 & \frac{1}{2}\Gamma_2 & 0 & \Gamma_3 \\ 0 & 0 & \frac{1}{2}\Gamma_2 & 0 & \Gamma_3 \\ 0 & 0 & \frac{1}{2}\Gamma_2 & 0 & -\Gamma_3 \end{bmatrix} \tag{6.31}$$

The matrices A_{pq}^i, B_{pq}^i, D_{pq}^i can be expressed as Eqs. (6.32)-(6.34), where $p, q \in 1,2,6$ and the superscripts have the same meaning with Eq. (6.18), i.e., i is the number of the tube, $\theta_{i,j}$ is the fiber winding angle of the *i*th tube's *j*-th layer.

$$A_{pq}^i = \sum_{j=1}^{N^{\mathrm{lay}}} (z_j - z_{j-1})\Gamma \begin{Bmatrix} 1 \\ \cos 2\theta_{i,j} \\ \sin 2\theta_{i,j} \\ \cos 4\theta_{i,j} \\ \sin 4\theta_{i,j} \end{Bmatrix}, p, q \in 1,2,6 \tag{6.32}$$

$$B_{pq}^i = \frac{1}{2}\sum_{j=1}^{N^{\mathrm{lay}}} (z_j^2 - z_{j-1}^2)\Gamma \begin{Bmatrix} 0 \\ \cos 2\theta_{i,j} \\ \sin 2\theta_{i,j} \\ \cos 4\theta_{i,j} \\ \sin 4\theta_{i,j} \end{Bmatrix}, p, q \in 1,2,6 \tag{6.33}$$

$$D_{pq}^{i}=\frac{1}{3}\sum_{j=1}^{\mathrm{N^{lay}}}(z_j^3-z_{j-1}^3)\Gamma\begin{Bmatrix}0\\ \cos 2\theta_{i,j}\\ \sin 2\theta_{i,j}\\ \cos 4\theta_{i,j}\\ \sin 4\theta_{i,j}\end{Bmatrix},p,q\in 1,2,6 \tag{6.34}$$

where A_{pq}^{i}, B_{pq}^{i}, D_{pq}^{i} are in-plane extensional stiffness, bending-stretching coupling stiffness and bending stiffness matrices, respectively.

The finite element method is used to analyze the response of the composite frame with classical laminate theory subjected to a given set of loading and boundary conditions. Blasques and Stolpe [162] had validated the analysis accuracy with beam, shell and solid finite element models to analysis composite beam. In the present research, the shell 99 element is adapted. The composite tube is divided into 8×200 meshes in ring and axial directions, respectively. The coupled degrees of freedom strategy at the end of the tube is utilized to realize the joints connecting of the composite tubes, and the joints can transfer moments and are assumed to be infinitely stiff. For the optimization in the case of continuous fiber winding angle, the elemental stiffness matrix of element n tube i, $\boldsymbol{K}^{i,n}$ is obtained as the integration of the constitutive matrix in the elements, as shown in Eq. (6.35). And the element constitutive matrix can be expressed as Eq. (6.36).

$$\boldsymbol{K}^{i,n}=\int_{\Omega^n}\boldsymbol{B}^{\mathrm{T}}\boldsymbol{D}_{\mathrm{c}}^{i}\boldsymbol{B}d\Omega^n \tag{6.35}$$

$$\boldsymbol{D}_{\mathrm{c}}^{i}=\begin{bmatrix}A_{pq}^{i}(\theta_{i,j},r_i) & B_{pq}^{i}(\theta_{i,j},r_i)\\ B_{pq}^{i}(\theta_{i,j},r_i) & D_{pq}^{i}(\theta_{i,j},r_i)\end{bmatrix} \tag{6.36}$$

In Eq. (6.35), $\boldsymbol{D}_{\mathrm{c}}^{i}$ is the elemental constitutive matrix of tube i, element n. The subscript c denotes the continuous micro fiber model. $\boldsymbol{B}$ is the strain–displacement matrix. The global stiffness matrix can be obtained as the sum of elemental stiffness over all $\mathrm{N^{ele}}$ elements, i.e., $\boldsymbol{K}=\sum_{n=1}^{\mathrm{N^{ele}}}\boldsymbol{K}^{i,n}$.

For discrete material model, the interpolation method must be implemented layer-wise for each element, i.e., for all layers in all elements. Then the elemental stiffness of A_{pq}^{i}, B_{pq}^{i} and D_{pq}^{i} should firstly be expressed as the sum of the candidate material stiffness similar as in Eq. (6.23) in this case. $A_{pq}^{i,j,k}$, $B_{pq}^{i,j,k}$, $D_{pq}^{i,j,k}$ is the stiffness of k-th candidate materials of the i-th tube's j-th layer. They can be obtained through Eqs. (6.32)-(6.34). Then we can obtain the stiffness expressions of n-th element of the i-th tube using a new discrete material model based on the HPDMO interpolation scheme as Eqs. (6.37)-(6.39).

$$A_{pq}^{i} = \sum_{j=1}^{\mathrm{N}^{\mathrm{lay}}} \sum_{k=1}^{\mathrm{N}^{\mathrm{can}}} \bar{w}_{i,j,k} A_{pq}^{i,j,k} \tag{6.37}$$

$$B_{pq}^{i} = \sum_{j=1}^{\mathrm{N}^{\mathrm{lay}}} \sum_{k=1}^{\mathrm{N}^{\mathrm{can}}} \bar{w}_{i,j,k} B_{pq}^{i,j,k} \tag{6.38}$$

$$D_{pq}^{i} = \sum_{j=1}^{\mathrm{N}^{\mathrm{lay}}} \sum_{k=1}^{\mathrm{N}^{\mathrm{can}}} \bar{w}_{i,j,k} D_{pq}^{i,j,k} \tag{6.39}$$

Then the structural constitutive matrix in discrete material model can be expressed as Eq. (6.40).

$$\boldsymbol{D}_{\mathrm{d}}^{i} = \begin{bmatrix} A_{pq}^{i} & B_{pq}^{i} \\ B_{pq}^{i} & D_{pq}^{i} \end{bmatrix} \tag{6.40}$$

where $\boldsymbol{D}_{\mathrm{d}}^{i}$ has the same meaning with $\boldsymbol{D}_{\mathrm{c}}^{i}$, just the subscript d denotes the discrete micro fiber model. Then we can solve the linear static equilibrium $\boldsymbol{KU} = \boldsymbol{F}$, where $\boldsymbol{U}$ and $\boldsymbol{F}$ are global vectors of the displacements and external forces of the composite structure, respectively.

(2) Sensitivity analysis of compliance optimization of composite frame

In order to perform gradient-based optimization efficiently, the design sensitivity analysis is done semi-analytically (Lund [163]; Cheng [164, 165]). In this section, two kinds of design variables, e.g., micro design variables $\theta_{i,j}$ (continuous fiber winding angle) or $x_{i,j,k}$ (discrete fiber winding angle) and macro design variable r_i are considered in two optimization models (DCMO and CCMO). Considering the limit of the pages of this section, this section only presents the compliance sensitivity analysis with respect to micro design variable $x_{i,j,k}$. The sensitivity of the compliance with respect to $\theta_{i,j}$ and r_i can be obtained in a similar procedure. Assumed that the applied static loads are designed independent, then the sensitivity of the objective function (i.e., the structural compliance C) in Eq. (6.19) with respect to the micro-scale design variable $x_{i,j,k}$ is given as Eq. (6.41).

$$\frac{\partial C}{\partial x_{i,j,k}} = \sum_{n=1}^{\mathrm{N^{ele}}} \left(\frac{\partial u_n^{\mathrm{T}}}{\partial x_{i,j,k}} \boldsymbol{K}_n \boldsymbol{U}_n + \boldsymbol{U}_n^{\mathrm{T}} \left(\frac{\partial K_n}{\partial x_{i,j,k}} \boldsymbol{U}_n + \boldsymbol{K}_n \frac{\partial \boldsymbol{U}_n}{\partial x_{i,j,k}} \right) \right) \tag{6.41}$$

where $\boldsymbol{U}_n$ are the displacement of element n, and $\boldsymbol{K}_n$ is the corresponding elemental stiffness of element n. Furthermore, after a further simplification, Eq. (6.41) can be simplified as Eq. (6.42).

$$\frac{\partial C}{\partial x_{i,j,k}} = \sum_{n=1}^{\mathrm{N^{ele}}} -\boldsymbol{U}_n^T \frac{\partial K_n}{\partial x_{i,j,k}} \boldsymbol{U}_n \tag{6.42}$$

$$\begin{aligned} \frac{\partial C}{\partial x_{i,j,k}} &= \sum_{n=1}^{\mathrm{N^{ele}}} -\boldsymbol{U}_n^{\mathrm{T}} \int_{\Omega^n} \boldsymbol{B}^{\mathrm{T}} \frac{\partial D_{\mathrm{d}}^i}{\partial x_{i,j,k}} \boldsymbol{B} d\Omega^n \boldsymbol{U}_n \\ &= \sum_{n=1}^{\mathrm{N^{ele}}} -\boldsymbol{U}_n^{\mathrm{T}} \int_{\Omega^n} \boldsymbol{B}^{\mathrm{T}} \begin{bmatrix} \frac{\partial A_{pq}^i}{\partial x_{i,j,k}} & \frac{\partial B_{pq}^i}{\partial x_{i,j,k}} \\ \frac{\partial B_{pq}^i}{\partial x_{i,j,k}} & \frac{\partial D_{pq}^i}{\partial x_{i,j,k}} \end{bmatrix} \boldsymbol{B} d\Omega^n \boldsymbol{U}_n \end{aligned} \tag{6.43}$$

It is possible to further extend the above equation with the elemental stiffness matrix defined in Eq. (6.35) and Eq. (6.40). Then the derivative of the compliance with respect to the design variable $x_{i,j,k}$ becomes as Eq. (6.43).

As we have mentioned, in the current implementation, the sensitivities $\partial \boldsymbol{D}_{\mathrm{d}}^i / \partial x_{i,j,k}$ are determined semi-analytically (Lund [163]; Cheng [164, 165]) with forward differences. Furthermore, the sensitivities of $\partial \boldsymbol{D}_{\mathrm{d}}^i / \partial x_{i,j,k}$ with respect to micro variables $x_{i,j,k}$ are implemented as follows

$$\frac{\partial D_{\mathrm{d}}^{i}(x_{i,j,k},r)_i}{\partial x_{i,j,k}} \approx \frac{D_{\mathrm{d}}^{i}\left((x_{i,j,k},r_i)+\mathrm{s}\cdot\boldsymbol{e}_k\right)-D_{\mathrm{d}}^{i}(x_{i,j,k},r)}{s} \tag{6.44}$$

where s is the step size and $\boldsymbol{e}_k$ is the unit vector of *k*-th candidate material. The step size s is set to 1×10^{-6} in our implementations.

The global volume constraint in Eq. (6.20) is only a function of the macro radius. The sensitivity of the structural volume constraint with respect to the inner radius r_i of the frame is easily obtained as

$$\frac{\partial v(r_i)}{\partial r_i} = \sum_{i=1}^{\mathrm{N^{tub}}} 2\pi\left[t_{\mathrm{tot}}^{i}+2r_i\right]\mathrm{L}_i \tag{6.45}$$

6.3.4 *Numerical examples and discussions*

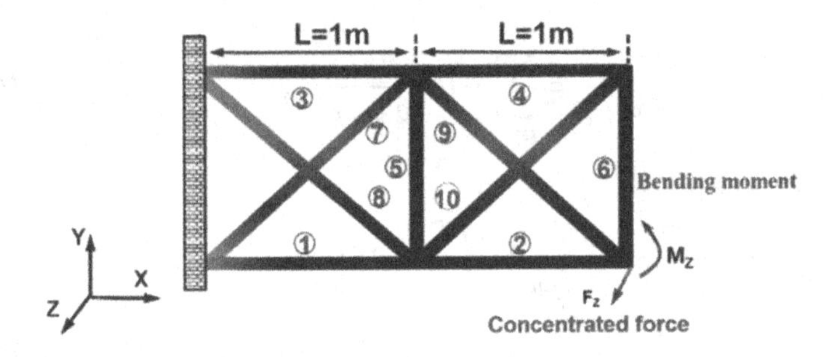

Fig. 6.18. A ten-beam composite frame

In this section, we consider a classical ten-beam composite frame as an academic example. Loading/boundary conditions and geometric sizes are shown in Fig. 6.18. The left part of the frame structure is fixed, and the right end corner of the structure is applied with an in-plane bending moment $\mathrm{M_Z} = 100\ \mathrm{kN\cdot m}$ and out-plane concentrated force $\mathrm{F_Z} = 100\ \mathrm{kN}$, respectively.

The fiber candidate materials are glass fiber reinforced epoxy with the orthotropic properties, $\mathrm{E_{11}} = 201\ \mathrm{GPa}$, $\mathrm{E_{22}} = 8.4\ \mathrm{GPa}$, $\mu_{12} = 0.25$, $\mathrm{G_{12}} = 4.2\ \mathrm{GPa}$, $\mathrm{G_{23}} = 2.1\ \mathrm{GPa}$. For easy of computation and no loss of generality, every tube is assumed to be composed with four winding layers, and the thickness of the circular tube is assumed with constant value $t_{\mathrm{tot}}^{i} = 0.4\ \mathrm{mm}$. The inner tube radius r_i are chosen as the macro-scale design variable, and the value range is $0 \leq r_{\min} \leq r_i \leq 100$ mm, $r_{\min}$ is the lower limit of the radius, which is set as $r_{\min} = 0.1$ mm in this section. For the discrete fiber winding angles, the most commonly used fiber winding angle $[0^\circ, +45^\circ, -45^\circ, 90^\circ]$ are adopted as the four candidate materials in model DCMO. The 90° winding angle means fiber is along the tube axial direction. Initial values of the micro-scale artificial density design variables are $x_{i,j,k} = 0.25$, i.e., the initial design of the candidate materials is uniformly distributed in each layer. In model CCMO, the fiber winding angle $\theta_{i,j}$ is directly considered as the microscale design variable. And the fiber winding angle range is $-90^\circ \leq \theta_{i,j} \leq 90^\circ$. Here, according to the numerical experience, we suggest using the $\theta_{i,j}=$ 14.3°, as the initial value of microscale design variables. In CCMOrad optimization model, the radius of components in the composite frame is optimized with the micro fiber winding angles fixed at $\theta_{i,j} = 90^\circ$. In the two optimization models, the initial value of macro-scale design variable is set as $r_i = 50$ mm. Then we can realize the topology optimization of frame structure through the macroscale optimization, and realize the microscale material optimization through the optimization of fiber winding angle.

In this numerical example, five kinds of optimization models have been investigated and compared. They are concurrent multiscale optimization models with consideration of continuous or discrete fiber winding angle and macrostructural radius as design variables, which can be denoted as CCMO and DCMO. Single-scale optimization models denoted as CCMOfib or CCMOrad are with either consideration of continuing fiber winding angle or macrostructural radius as design variables, respectively. And we also defined the single scale optimization denoted as DCMOfib in which only the discrete micro material optimization is carried out.

Table 6.3 shows the initial value of macro- and micro-variables and the comparison of the optimized compliance from the above five models, where $x_{i,j,k}$ is the artificial candidate material density. Table 6.4 gives the comparison of the concurrent optimized results of DCMO and CCMO models, respectively. And Fig. 6.19 gives the iteration history of the objective function. It can be directly observed from Table 6.3 and Fig. 6.19 that the multiscale optimization results from DCMO are obviously better than the single-scale optimization results CCMOrad, DCMOfib, and CCMOfib, with decrease of 41.94%, 33.98%, and 32.63%, respectively. It fully reflects the advantages of concurrent multiscale optimization of the composite frame. It also can give a reasonable explanation from the point view of a wider design domain in the concurrent optimization. The concurrent multiscale optimization sufficiently takes into account the coupling effects of the macrostructure and micro-material to maximize the potential of composite frame structures. From the comparison with other three single-scale optimization models in the optimization iteration history of Fig. 6.19, DCMO model can give out better optimization results with the same or less iteration steps. And the convergence rate in Table 6.3 shows that the micro fiber winding angles are completely convergent in DCMO model with HPDMO interpolation scheme.

Table 6.3. Comparison of optimized results of a ten-beam composite frame structure

Initial micro design variables	Initial macro variables	Five optimization models	Objective Function	Objective function compared with CCMOrad/%	Convergence rate $H_\xi = \frac{N_c^{l,tol}}{N^{l,tol}}/\%$
90°	0.05	CCMOrad	233608.53	0%	--
$x_{i,j,k} = 0.25$	0.05	DCMOfib	215737.97	7.65%	100%
14.3°	0.05	CCMOfib	201324.62	13.82%	--
$x_{i,j,k} = 0.25$	0.05	DCMO	135617.20	41.94%	100%
14.3°	0.05	CCMO	121204.87	48.11%	--

Note: $x_{i,j,k}$ is the artificial candidate material density

While if we compare the CCMO and the DCMO, we can find that the structural compliance from CCMO model decreases further by 10.63%, i.e., the averaged structural stiffness is increased by 10.63%. This is because the candidate winding fiber angle are limited within $[0^\circ, \mp 45^\circ, 90^\circ]$ in DCMO model. This actually forced a constraint on the design variables, thus to reduce the design optimization space of DCMO model at the microscale and lead to a relatively higher structural compliance. However, it should be especially pointed out that the optimized result from the DCMO model can be manufactured with a relatively lower cost.

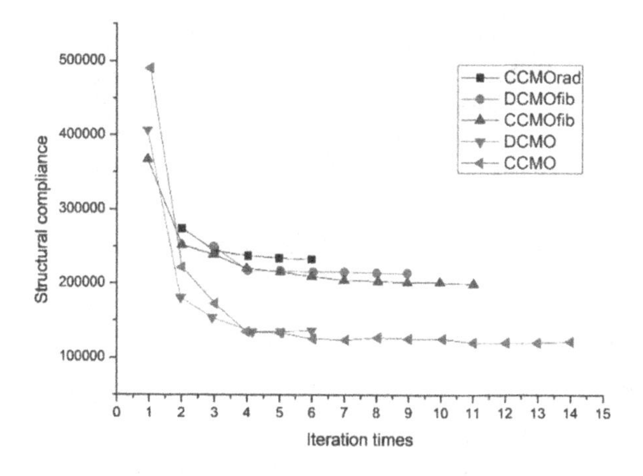

Fig. 6.19. Iteration history of the objective function

Furthermore, observed from the following Fig. 6.20(a) and (b), the multiscale optimization models DCMO and CCMO can give the consistent optimized configurations in macroscale. For example, the tubes 7, 9, 10 have nearly reach the lower limit of the macro design variable and can be regarded to be deleted from the original ground structure. In the present research, the joints connecting the composite tubes can transfer moments and are assumed to be infinitely stiff. It should be pointed out that, in the DCMO model, the micro fiber winding angles are restricted in a specific candidate angle set, which led to the macro- and micro-design variables can't be freely coupled, and thus the tube's number 7, 9, 10 don't fully reach the lower limit (however, much lower than the cross-sectional areas of other tubes in the optimized structure), the detail of the macroscale radius can be found in Table 6.4.

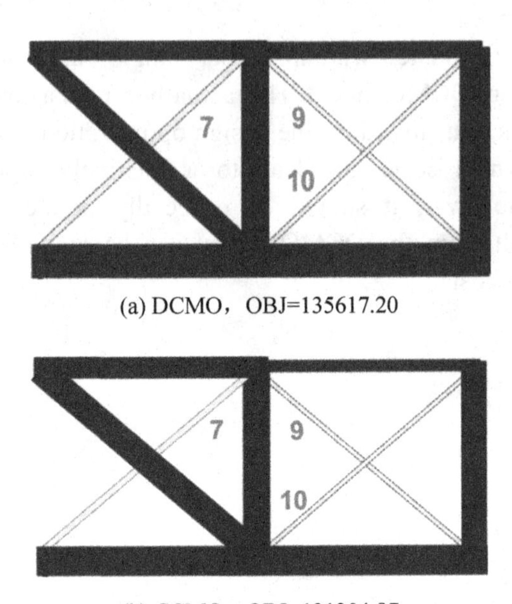

(a) DCMO，OBJ=135617.20

(b) CCMO，OBJ=121204.87

Fig. 6.20. Optimized configurations and objective functions of DCMO and CCMO model

Then from the point view of the micro fiber winding angle, the tubes of the optimized composite frame can be divided into 4 groups denoted as G1-G4 with different representative colors. The G1 group contains tubes 1 and 3, the G2 group has tubes 2, 4, 7, 8, 10, the G3 group contains tubes 5, 6, and the G4 group contains tube 9.

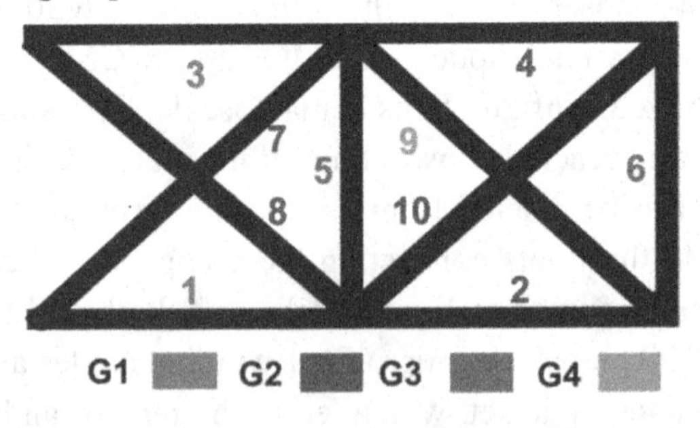

Fig. 6.21. Illustration of the group classification

The tubes in the same group share the similar micro fiber winding properties in the two concurrent optimization models. Observing from Table 6.4, the fiber winding angles of the tubes in G1 from DCMO model are all 90°, and the optimized winding angles of the inner and outer layers in G1 are 89.62°(tube 1), 88.58°(tube 1), 89.89° (tube 3), 89.47°(tube 3) in CCMO model, which can be rounded to approximately 90°. The fiber winding angles of the middle two layers in G1 are 74.91° (tube 1), 72.47° (tube 1), 70.56°(tube 3), 71.78° (tube 3), which are close to the 90° in CCMO model. G2 and G3 also have the similar observations. Only G4 (i.e., tube 9) does not have this similarity. But this tube has achieved the lower limit of the macro design variable which means it is actually deleted from the initial ground structure and has little influence on the overall structural stiffness. So, from the above analysis, we can see that the optimized fiber winding angle in DCMO and CCMO models is almost consistent, which can verify the correctness and reliability of the DCMO model and its optimized results in some extends.

Table 6.4. Concurrent multiscale optimized results of DCMO and CCMO model

Tube	Macro variables/m		Micro design variables/°							
			1st layer		2nd layer		3rd layer		4th layer	
	DCMO	CCMO	DCMO	CCMO	DCMO	CCMO	DCMO	CCMO	DCMO	CCMO
1	0.0940	0.0998	90	89.62	90	74.91	90	72.47	90	88.58
2	0.0882	0.0941	90	86.77	45	48.23	−45	52.74	90	89.92
3	0.0456	0.0450	90	89.89	90	70.56	90	71.78	90	89.47
4	0.0484	0.0210	90	87.50	−45	47.53	45	52.68	90	87.90
5	0.0940	0.0998	45	54.75	−45	45.77	45	−44.13	−45	53.66
6	0.0840	0.0998	−45	52.76	−45	8.55	45	−46.99	45	68.75
7	0.0068	0.0001	90	83.71	−45	62.86	45	63.22	90	83.65
8	0.0591	0.0868	90	89.93	−45	70.57	45	67.91	90	89.05
9	0.0076	0.0001	90	89.05	−45	89.05	45	89.05	90	89.05
10	0.0071	0.0001	90	87.16	−45	69.99	45	71.90	90	87.02

When we compared the optimization results from Fig. 6.20(a) and Fig. 6.22, we can find the structural compliance from the DCMO model is much lower than that from the CCMOrad model. And the configurations of macro structure are different. In CCMOrad optimization model, the radius of components in the composite frame is optimized with the micro fiber winding angles fixed at $\theta_{i,j} = 90^{\circ}$. Then we can conclude that the optimized macro configuration of the CCMOrad will be different with the different fixed micro fiber winding angles. But it is impossible for engineers to give the optimal initial fiber winding angles for a complex composite frame before the optimization. So, this actually reflects the significance of the concurrent multiscale optimization of the composite frame, which fully considers the coupling effect of the structural configuration and material design, there by saves lots of design iterations and computational resources. And from the comparison of Fig. 6.20(a) and Fig. 6.22, Tubes 7, 10 have reached the lower limit of macro design variable in both the figures, while the cross-sectional area is different from tube 5. We think the configuration of Fig. 6.20(a) is more consistent with the loading characteristics of bending moment M_Z and concentrated force F_Z, which can partly verify the effectiveness of DCMO optimization model.

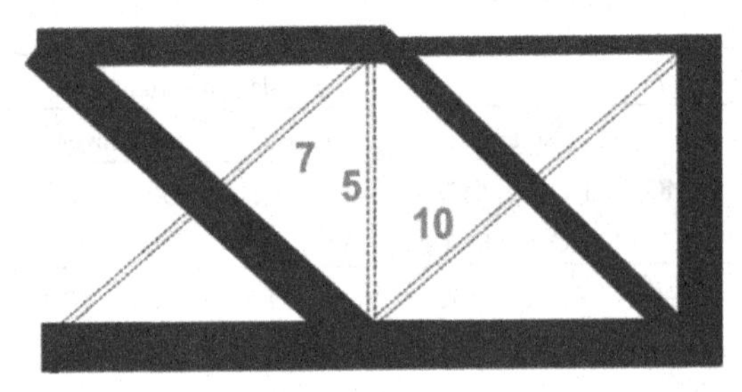

Fig. 6.22. Optimized configuration and objective function of CCMOrad, OBJ=233608.53

Table 6.5. Macro-scale optimized results of CCMOrad model

Beam number	1	2	3	4	5
Radius/m	0.0999	0.0999	0.0625	0.0352	0.0001
Beam number	6	7	8	9	10
Radius/m	0.0999	0.0001	0.0984	0.0327	0.0001

Table 6.5 gives out the results of the single scale optimization of the fiber winding angle from DCMOfib and CCMOfib. Because only the fiber winding angle is considered as the design variable, the structural configuration is fixed as show in Fig. 6.23(a) and Fig. 6.23(b). The macro radius is kept the constant value $r_i = 0.05$ m in the optimization procedure. The tubes (1-3,10) and (4-9) are classified into different groups used in the following discussions.

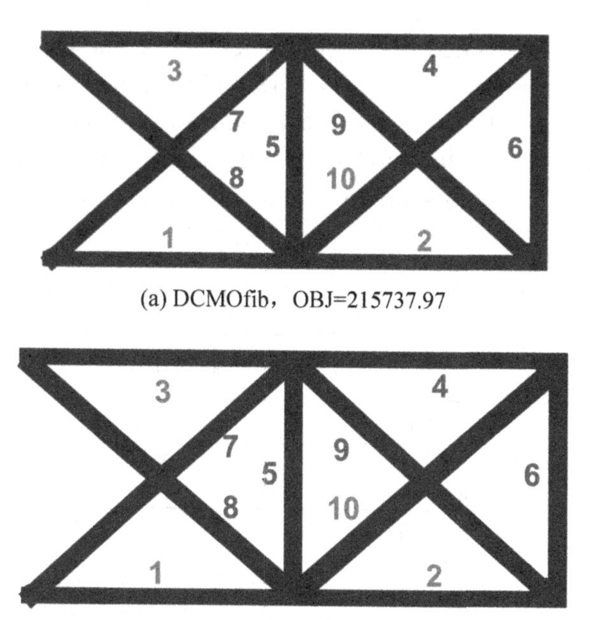

(a) DCMOfib, OBJ=215737.97

(b)CCMOfib, OBJ=201324.62

Fig. 6.23. Optimized configurations and objective functions of DCMOfib and CCMOfib models

From the optimization results of Table 6.6 we can find that the optimized fiber winding angle from DCMOfib model is 90° in tubes 1, 2, 3, 10, and the optimized angle is a combination with the four candidates for the other six tubes (4, 5, 6, 7, 8, 9). Observing the structural loading conditions shown in Fig. 6.18, tubes 1, 2, 3, 10 mainly bear the tension or compression along their axial of the beam caused by the applied concentrated forces and bending moment. Thus, the filament along the axial i.e., 90° fiber winding angle is the best choice to transfer the axial loads. However, the tubes 4, 7, 8, 9 not only bear the axial force, but also the shear force. Thus, the outer and inner layers with 90° fiber winding angle can effectively resist the axial loads, while the middle layer with $\mp 45^{\circ}$ fiber winding angle can effectively resist the deformation caused by shear force, then lead to a greater contribution to the overall stiffness than that from the angle of 90°. Tubes 5, 6 are with the combination of $\mp 45^{\circ}$ fiber winding angles because tubes 5, 6 mainly bear to the shear loads, which is consistent with the optimized angle designs.

It should be pointed that, as discussed above, the CCMOfib model has larger design space compared with DCMOfib model in the microscale. So CCMOfib model can obtain better optimization results than DCMOfib model. And for CCMOfib model, the optimized fiber winding angles of tubes 1, 2, 3, 10 are close to 90° after rounding same as the design of DCMOfib.

Table 6.6. Optimized results of DCMOfib and CCMOfib models

Tube	Macro variables/m		Micro design variables							
			First layer		Second layer		Third layer		Fourth layer	
	DCM Ofib	CCM Ofib	DCM Ofib	CCM Ofib	DCM Ofib	CCM Ofib	DCM Ofib	CCM Ofib	DCM Ofib	CCM Ofib
1	0.05	0.05	90	89.74	90	73.52	90°	72.31	90	89.31
2	0.05	0.05	90	87.63	90	88.78	90	89.59	90	89.55
3	0.05	0.05	90	89.48	90	85.56	90	86.34	90	89.38
4	0.05	0.05	90	89.75	-45	46.32	45	49.76	90	89.56
5	0.05	0.05	-45	−44.83	45	45.39	45	44.24	45	46.35
6	0.05	0.05	45	56.36	45	−45.34	-45	−46.77	45	52.56
7	0.05	0.05	90	88.92	-45	58.59	45	61.24	90	89.34
8	0.05	0.05	90	89.72	45	68.47	45	67.78	90	89.59
9	0.05	0.05	90	89.74	45	45.33	45	46.57	90	89.28
10	0.05	0.05	90	89.17	90	79.85	90	79.83	90	88.96

6.4 *Optimization design with maximizing fundamental frequency*

6.4.1 *Mathematical formula of maximizing the fundamental frequency*

In the section, we consider the concurrent multiscale optimization of the composites frame with the objective of maximizing the fundamental frequency under total volume constraints. The mathematical formula of the optimization problem can be expressed as follows.

$$\text{find} \quad \boldsymbol{X} = \{r_i, \theta_{i,j}\} \tag{6.46}$$

$$\text{max} \quad f = \omega_1^2\{r_i, \theta_{i,j}\} \tag{6.47}$$

$$\begin{aligned} \text{s.t.} \quad & \boldsymbol{K\phi}_k = \omega_k^2 \boldsymbol{M\phi}_K \\ & V(r_i) = \sum_{i=1}^{\mathrm{N^{tub}}} \pi\left[t_{\mathrm{tot}}^i \times t_{\mathrm{tot}}^i + 2r_i t_{\mathrm{tot}}^i\right]\mathrm{L}_i \leq \bar{V} \\ & -90^\circ \leq \theta_{i,j} \leq 90^\circ \\ & r_{\min} \leq r_i \leq 0.2 \\ & i \in \mathrm{N^{Tub}}, j \in \mathrm{N^{Lay}} \end{aligned} \tag{6.48}$$

where $f = \omega_1^2(r_i, \theta_{ij})$ is the objective function, ω_1 is the fundamental natural frequency. The macro-scale design variable r_i is the inner radius of the composite tubes with circular section. The microscale design variable $\theta_{i,j}$ is the continuous fiber winding angle, where the subscript i and j denote the number of tubes and layers respectively. ω_k is the k-th order frequency and $\boldsymbol{\phi}_k$ is the corresponding eigenvector, and $\boldsymbol{K}$ and $\boldsymbol{M}$ are the symmetric and positive definite stiffness and mass matrix. $\mathrm{N^{tub}}$ and $\mathrm{N^{lay}}$ denote the total number of tubes and layers. $\bar{V}$ is the entire volume of the macro-design domain. $r_{\min}$ ($r_{\min} = 0.0001\mathrm{m}$ in present section) is a small positive value to avoid the singularity during the optimization iterations. t_{tot}^i is the total layer thickness of the ith composite tube. In the macroscale, the theory of Ground Structure Method [141, 166] is adopted to realize the topology optimization when the radius reaches its lower limit $r_{\min}$.

6.4.2 *Sensitivity analysis*

In order to perform gradient-based optimization efficiently, the work adopts the semi-analytical method (SAM) [162-165] which has highly computational efficiency and is widely used in the sensitivity analysis of finite element models. This section only presents the fundamental frequency sensitivity analysis with respect to the micro design variable $\theta_{i,j}$. The sensitivity about the macroscale design variable r_i can be obtained in a similar procedure. The direct approach to obtain the eigenvalue sensitivities is to differentiate the generalized vibration eigenvalue equation without damping $\boldsymbol{K\phi}_k = \omega_k^2 \boldsymbol{M\phi}_k$ with respect to the design variable $\theta_{i,j}$ as given as Eq. (6.49).

$$\frac{\partial \boldsymbol{K}}{\partial \theta_{i,j}}\boldsymbol{\phi}_k + \boldsymbol{K}\frac{\partial \boldsymbol{\phi}_k}{\partial \theta_{i,j}} = \frac{\partial \omega_k^2}{\partial \theta_{i,j}}\boldsymbol{M\phi}_k + \omega_k^2 \frac{\partial \boldsymbol{M}}{\partial \theta_{i,j}}\boldsymbol{\phi}_k + \omega_k^2 \boldsymbol{M}\frac{\partial \boldsymbol{\phi}_k}{\partial \theta_{i,j}} \tag{6.49}$$

Pre-multiplying Eq. (6.49) by $\boldsymbol{\Phi}_k^{\mathrm{T}}$, we can get Eq. (6.50).

$$\begin{aligned} &\boldsymbol{\phi}_k^{\mathrm{T}}\frac{\partial \boldsymbol{K}}{\partial \theta_{i,j}}\boldsymbol{\phi}_k + \boldsymbol{\phi}_k^{\mathrm{T}}\boldsymbol{K}\frac{\partial \boldsymbol{\phi}_k}{\partial \theta_{i,j}} \\ &= \boldsymbol{\phi}_k^{\mathrm{T}}\boldsymbol{M\phi}_k\frac{\partial \omega_k^2}{\partial \theta_{i,j}} + \omega_k^2\boldsymbol{\phi}_k^{\mathrm{T}}\frac{\partial \boldsymbol{M}}{\partial \theta_{i,j}}\boldsymbol{\phi}_k + \omega_k^2\boldsymbol{\phi}_k^{\mathrm{T}}\boldsymbol{M}\frac{\partial \boldsymbol{\phi}_k}{\partial \theta_{i,j}} \end{aligned} \tag{6.50}$$

Pre-multiplying $\boldsymbol{K\phi}_k = \omega_k^2 \boldsymbol{M\phi}_k$ by $\left(\partial \boldsymbol{\phi}_k / \partial \theta_{i,j}\right)^{\mathrm{T}}$, we can obtain Eq. (6.51).

$$\left(\frac{\partial \boldsymbol{\phi}_k}{\partial \theta_{i,j}}\right)^{\mathrm{T}} \boldsymbol{K\phi}_k = \omega_k^2 \left(\frac{\partial \boldsymbol{\phi}_k}{\partial \theta_{i,j}}\right)^{\mathrm{T}} \boldsymbol{M}\phi_k \tag{6.51}$$

Then the sensitivity of the fundamental frequency with respect to microscale design variables can be further expressed as Eq. (6.52) by substituting Eq. (6.51) into Eq. (6.50).

$$\frac{\partial \omega_k^2}{\partial \theta_{i,j}} = \frac{\boldsymbol{\phi}_k^{\mathrm{T}}\frac{\partial \boldsymbol{K}}{\partial \theta_{i,j}}\boldsymbol{\phi}_k - \omega_k^2\boldsymbol{\phi}_k^{\mathrm{T}}\frac{\partial \boldsymbol{M}}{\partial \theta_{i,j}}\boldsymbol{\phi}_k}{\phi_k^T \boldsymbol{M}\phi_k} \tag{6.52}$$

Adopt the mass-normalized mode shape $\widehat{\boldsymbol{\phi}}_k$, which satisfies $\widehat{\boldsymbol{\phi}}_k{}^{\mathrm{T}}\boldsymbol{M}\widehat{\boldsymbol{\phi}}_k = 1$, then Eq. (6.52) can be written as Eq. (6.53).

$$\frac{\partial \omega_k^2}{\partial \theta_{i,j}} = \widehat{\boldsymbol{\phi}}_k^{\mathrm{T}}\frac{\partial \boldsymbol{K}}{\partial \theta_{i,j}}\widehat{\boldsymbol{\phi}}_k - \omega_k^2\widehat{\boldsymbol{\phi}}_k^{\mathrm{T}}\frac{\partial \boldsymbol{M}}{\partial \theta_{i,j}}\widehat{\boldsymbol{\phi}}_k \tag{6.53}$$

In the current implementation, the sensitivities $\partial \boldsymbol{K}/\partial \theta_{i,j}$ and $\partial \boldsymbol{M}/\partial \theta_{i,j}$ are determined by semi-analytical forward differences. The approach has higher computational efficiency than the OFD (overall finite difference) method, because the computation of the frequency equation with global stiffness and mass matrices, which is the most time-consuming part in the optimization, will be only calculated once for N design variables. Contrarily, in the OFD method, the frequency equation needs to be calculated at least $N+1$ times for N design variables. The sensitivities of $\boldsymbol{K}$ and $\boldsymbol{M}$ with respect to micro variables $\theta_{i,j}$ are expressed as follows.

$$\frac{\partial \boldsymbol{K}(\theta_{i,j})}{\partial \theta_{i,j}} \approx \frac{\boldsymbol{K}\left((\theta_{i,j}) + \mathrm{s}\cdot\theta_{i,j}\right) - \boldsymbol{K}(\theta_{i,j})}{\mathrm{s}}$$
$$\frac{\partial \boldsymbol{M}(\theta_{i,j})}{\partial \theta_{i,j}} \approx \frac{\boldsymbol{M}\left((\theta_{i,j}) + \mathrm{s}\cdot\theta_{i,j}\right) - \boldsymbol{M}(\theta_{i,j})}{\mathrm{s}} \tag{6.54}$$

where s represents the step size and has been set to 1×10^{-6} in this implementation. Then the sensitivity of the fundamental frequency with respect to r_i can be obtained in a similar procedure as $\theta_{i,j}$.

The global volume constraint in Eq. (6.48) is only a function of the macro radius design variables. The sensitivity of the volume with respect to the radius r_i of the frame can be easily obtained as

$$\frac{\partial V(r_i)}{\partial r_i} = 2\pi r_i \frac{t_{\text{tot}}^i}{r_i^0}\left(\frac{t_{\text{tot}}^i}{r_i^0} + 2\right) L_i \tag{6.55}$$

where t_{tot}^i and r_i^0 are the original total layer thickness and inner radius, and $\mathrm{L_i}$ denotes the length of the *i*th tube. In this work, the layer thickness is proportionate to the corresponding tube's radius as dependent variables.

6.4.3 *Numerical examples and discussions*

To confirm the effectiveness of the concurrent multiscale optimization model of composite frame structures with maximum fundamental frequency proposed in present section, this section shows two numerical examples about plan ten-tube structure and spatial sixteen-tube structure. The previous Section 6.3.4 had detailed descriptions about the single- and multiscale optimization of composite material structures, and pointed out the advantage of the concurrent multiscale optimization design. Therefore, the present section will only present the concurrent multiscale optimization results of composite frame structures with maximum fundamental frequency.

(1) **Example 1:** Planar ten-tube composite frame

The loading/boundary conditions and geometric sizes of the planar ten-tube composite frame are shown in Fig. 6.24. In the section, all the physical quantities are dimensionless. The marks in the picture are concentrated mass on the nodes which have the magnitude of 100. The left side clamped, and the tube length L = 1. Adopting the two-scale optimization model like Eqs. (6.46)-(6.48), we achieve the concurrent multiscale optimization of composite frame with maximum fundamental frequency under the material volume constraint.

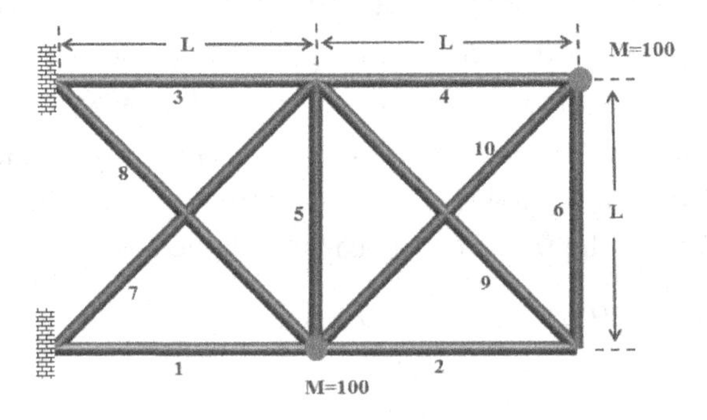

Fig. 6.24. A ten-tube composite frame

The orthotropic properties of the fiber reinforced composites are as follows: $E_{11} = 1.43 \times 10^{11}$, $E_{22} = 10^9$, $\mu_{12} = 0.25$, $G_{12} = 6 \times 10^9$, $G_{23} = 5 \times 10^9$, $\rho = 2900$. The cross-sectional shape of the beam is circular, which is the conventional structural form in the practical engineering. Each tube is assumed to be composed of ten winding layers. The initial layer thickness and inner radius of the circular tube are assumed as $t_0 = 0.008$, $r_0 = 0.2$, respectively. The inner tube radius r_i are chosen as the macroscale design variables, and the value range is $0 \leq r_{\min} \leq r_i \leq 0.2$, where $r_{\min}$ is the lower limit of the radius, which is set as $r_{\min} = 0.0001$ in this section. And the fiber winding angles $\theta_{i,j}$ are chosen as the microscale design variables, and the value range is $-90^\circ \leq \theta_{i,j} \leq 90^\circ$, with the initial angle $\theta_{i,j} = 45°$. The 0° winding angle means fiber is along the tube's axial direction. The volume constraint is the total amount of the composite material, $V(r_i) = \sum_{n=1}^{\mathrm{NTub}} v_n \leq \bar{V}$, and $\bar{V}$ is the upper limit which is taken as the initial structural volume.

Considering the realization of the topology optimization and the harmonious relations about the radius and thickness of the tube in the practical engineering, in this section, we adopt a punitive relationship between the winding layer thickness and the tube's inner radius as follows.

$$\tilde{t}_i = (r_i/r_0)^{\alpha} \times t_0, i \in \mathrm{N}^{\mathrm{Tub}} \ if \ r_i < r_0 \tag{6.56}$$

where $\tilde{t}_i$ is the layer thickness of tube i after the punishment, r_i is the inner radius of tube i in the current iteration. $r_0 = 0.02$ is the initial inner radius of the tube. α is the penalization index and we set $\alpha = 1$ in the section. Through the penalization relationship, the layer thickness change with the inner radius of each tube when the inner radius is lower than initial value, so that we can establish an adaptive process within macro radius and layer thickness. Based on the ground structure approach of topology optimization, the decrease of the macro radius proves that the tube has lower contribution to the stiffness of the whole structure, so that we can give more material to the other tubes to improve the stiffness by decreasing the layer thickness of the weak one. When the macro radius is close to the lower limit, the layer thickness has been small enough to be deleted. In this way, we can not only achieve the topology optimization, but also reduce the amount of calculation considering the thickness to improve the optimization iterative efficiency.

Table 6.7 shows the concurrent optimization results of ten-tube frame with the macro variables r_i and the micro variables $\theta_{i,j}$. Fig. 6.25 show the iteration history, optimized topology structure and the first order vibration mode of ten-tube structure, respectively. As shown in the iteration history of the structural fundamental frequency in Fig. 6.25, the structural fundamental frequency enhances 340.97% compared to the initial value of the structure and improved the dynamic performance of the structure significantly. Certainly, the enhancement of the fundamental frequency is based on the initial variable value, but for complex concurrent optimization studied in the section, it is also difficult to give a better initial value for an engineer from only engineering experience. Adopting the homogeneous initial value is a general method in engineering structural optimization.

Table 6.7. Concurrent optimization results of the planar ten-tube composite frame

Tube number	Optimized macro variables r_i	Optimized fiber winding angle/ °
1	0.1136	0/0/0/0/0/0/0/0/0/0
2	0.0067	-11/-14/-13/-9/-4/-4/0/0/0/0
3	0.0964	0/0/0/0/0/0/0/0/0/0
4	0.0186	0/0/0/0/0/0/0/0/0/0
5	0.0076	0/0/0/0/0/0/0/0/0/0
6	0.0024	0/0/0/0/0/0/0/0/0/0
7	0.0279	0/0/0/0/0/0/0/0/0/0
8	0.0897	0/0/0/0/0/0/0/0/0/0
9	$r_{\min}$	-14/21/24/25/26/-26/-24/19/-1/0
10	0.0162	0/0/0/0/0/0/0/0/0/0
Optimized fundamental frequency ω_1		242.8440

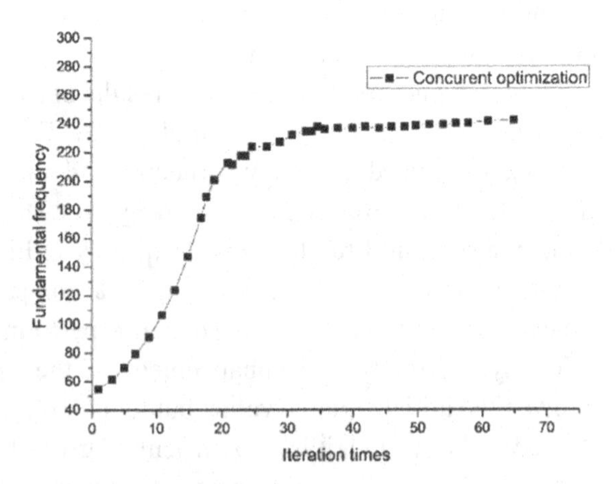

Fig. 6.25. The structural fundamental frequency iteration history of the planar ten-tube frame

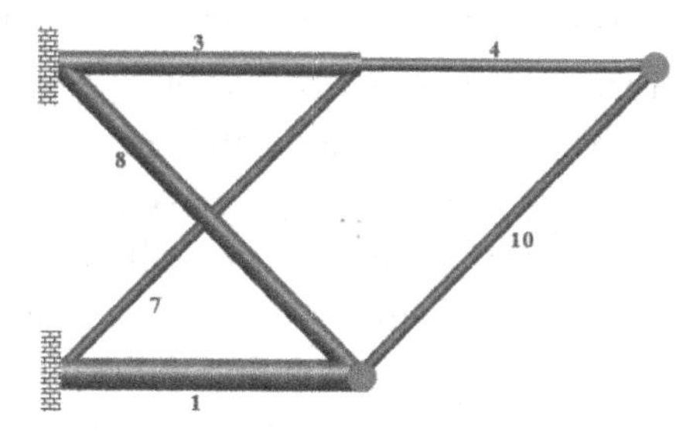

Fig. 6.26. The optimized topology structure of the ten-tube composite frame

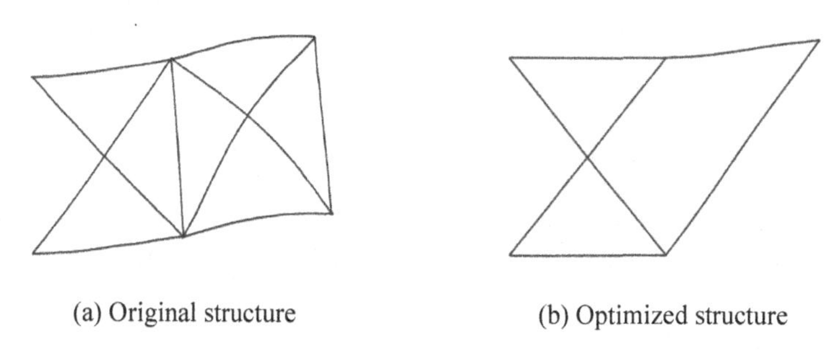
(a) Original structure (b) Optimized structure

Fig. 6.27. The first order vibration mode of the ten-tube frame

Table 6.7 gives the result of the concurrent optimization. The macro design variables of four tubes have reached or been close to the lower limit. Among these tubes, the number 9 tube has reached the lower limit. And the tubes 2, 5, 6 are small enough compared with the other radius of the tubes. Through the penalization relationship of Eq. (6.56), the thickness of these four tubes can be ignored. These tubes have lower contribution to the stiffness of the whole structure so that we can delete these tubes from the ground structure. Fig. 6.26 gives the macro topology optimized structure, where the tubes 2, 5, 6, 9 have been deleted, and get a six-tube structure at the end. The horizontal tubes 1, 3 near the clamped point and the tube 8 which connects the under concentrated mass with the clamped point are thicker than the others to enhance the structural stiffness in a limited total amount of material. The result corresponds with the best transmission path of mechanical obviously.

It can be observed from the result of the micro design variables $\theta_{i,j}$ (left hand is the inner layer) that all the fiber winding angles are 0° except for tubes 2, 9. But as the discussions above, the macro variables of the two tubes have achieved the lower limit and have lower contribution on stiffness to the whole structure. The result actually reflects the structural form and loading/boundary conditions of the example. The first order vibration mode of planar structure is swinging in the plane as Fig. 6.27, so that the fibers winding along the axis of the tubes can improve the stiffness of the structure. However, with regard to the spatial frame, the vibration mode contains not only bending but also torsion, for which the helical winding fibers will help to improve the shearing and torsion stiffness. The following examples will give a further explanation.

(2) **Example 2:** Spatial sixteen-tube composite frame

The Loading/boundary conditions and geometric sizes of a spatial sixteen-tube composite frame are shown in Fig. 6.28. The concentrated mass on the nodes have the magnitude of 30 respectively. The bottom part of the structure is fixed.

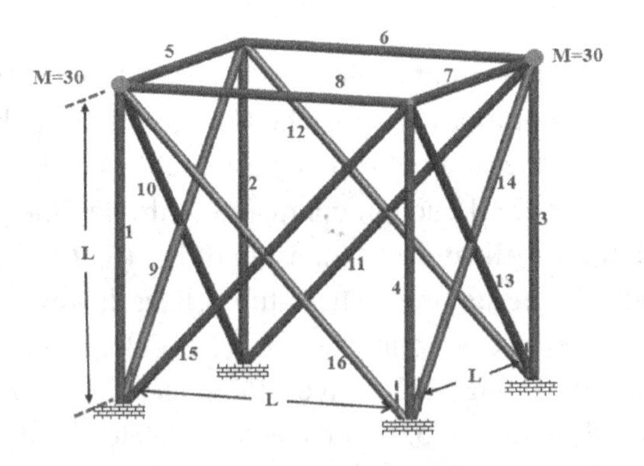

Fig. 6.28. The example of a sixteen-tube composite frame

The tubes are consisting of the fiber reinforced composite material, all the material properties and the parameters the same as example 1. Fig. 6.29-Fig. 6.31 show the iteration history, optimized topology structure configuration and the first order vibration mode of the sixteen-tube structure, respectively. From Fig. 6.29, we can find that the fundamental frequency enhances 152.50% compared with the initial fundamental frequency of the structure, which significantly improved the structural dynamic performance like example 1. Table 6.8 shows that tubes 5-8 reach the lower limit and we can delete these tubes from the ground structure. As shown in the macro optimized topology structure in Fig. 6.30, the tubes 1-4 have a relatiely larger inner radius, and tubes 1, 3 which connected to the concentrated mass are thicker than the other tubes. It demonstrates that giving more material to the four erect tubes will lead to a great contribution to the overall stiffness especially the tubes connected to the lumped mass. While other slanting tubes are relatively thinner under the constraint of material volume and assemble all the tubes together.

Table 6.8. Concurrent optimization results of the sixteen-tube composite frame

Tube number	Optimized macro variablesr_i	Optimized fiber winding angle/ °
1	0.0578	-3/-4/-4/-4/0/0/0/0/0/0
2	0.0531	-6/-6/-6/-4/-5/-4/0/0/0/0
3	0.0655	-2/-3/-3/-2/-2/1/1/2/1/1
4	0.0456	-7/-8/-8/-9/-4/-4/1/1/1/-2
5	0.0006	-2/14/19/16/14/11/-5/0/0/0
6	$r_{\min}$	11/16/21/22/21/15/-5/0/1/1
7	$r_{\min}$	13/19/24/24/24/24/21/-10/0/0
8	0.0005	-5/17/21/21/21/16/-5/-1/-1/1
9	0.0208	-6/-12/-12/-11/-6/-5/-3/-3/-2/2
10	0.0247	9/14/15/14/8/4/0/0/0/0
11	0.0219	-3/10/11/-11/-7/-3/-1/0/0/-1
12	0.0318	12/15/16/14/10/4/0/-1/-1/-2
13	0.0225	-9/-15/-16/15/-10/-4/-2/-1/1/1
14	0.0257	12/12/16/11/10/3/2/2/0/-1
15	0.0207	-6/13/-15/-14/-10/-5/-4/0/-1/-1
16	0.0212	2/12/12/12/7/3/2/1/1/0
Optimized fundamental frequencyω_1		296.1600

The micro variables of the optimized results in Table 6.8 show that all the tubes in the structure have non-zero winding angles which is different from the planar structure. According to Fig. 6.31 the first order vibration mode of the sixteen-tube structure is more complex than the planar one. Besides the vibration of the whole structure, each tube of the structure will vibrate in different directions so that more bending and torsion deformation occur. To resist the shear force caused by bending and torsion deformation, the non-zero winding angle will contribute more to the overall stiffness than that from angle 0°. Winding angles in different layers have different effects on the structure's stiffness. For example, most of the outer winding angles are approximate to 0°, which improve the bending stiffness of the structure, and all of the inner winding angles are non-zero, which improve the torsion stiffness. Meanwhile, the optimization fully takes into account the coupling effect of macro- and micro-variables. It should be pointed out that most of the composite frame structures are in the form of space, so it is extremely important to investigate the effect of micro fiber winding angles on the structural stiffness and dynamic frequency performance.

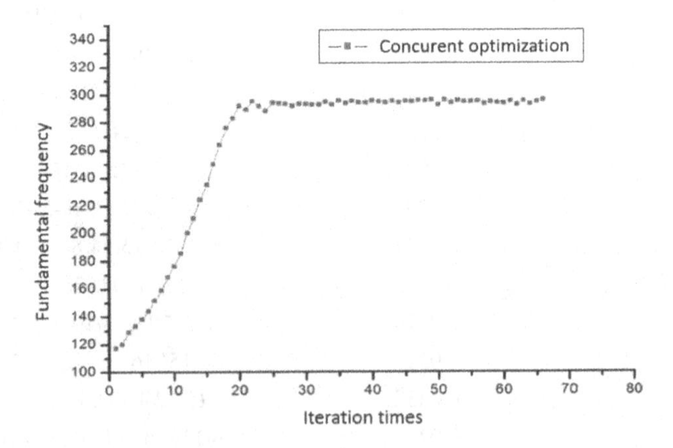

Fig. 6.29. The structural fundamental frequency iteration history of the spatial sixteen-tube frame

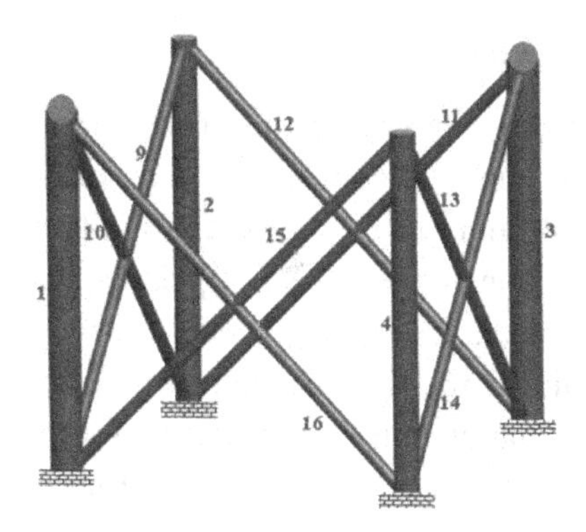

Fig. 6.30. The optimized topology structure of the sixteen-tube composite frame

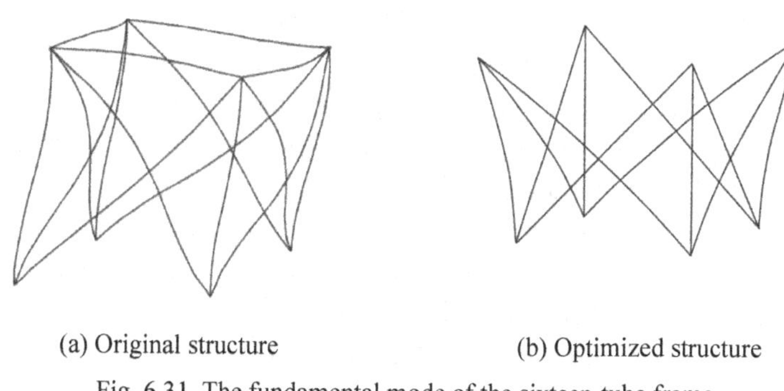

(a) Original structure (b) Optimized structure

Fig. 6.31. The fundamental mode of the sixteen-tube frame

6.5 *Optimization design with specific manufacturing constraints*

At the microscopic level, the discrete fiber winding angle selection is solved by using the HPDMO approach in the sense that the structural constituents are chosen from among a given set of candidate materials. In most practical applications, the candidate composite ply is restricted to $\left[0^{\circ},+45^{\circ},-45^{\circ},90^{\circ}\right]$, which are the conventional orientations used in aeronautics [159]. Considering the manufacturing process, Mallick [167] suggested that 0° and 90° fiber winding angles in filament winding process should be implemented by 5° and 85° fiber winding angles, respectively. So, in this work, we consider the assembly of $\left[5^{\circ},+45^{\circ},-45^{\circ},85^{\circ}\right]$ as a set of candidate composite fiber winding angles. The fiber winding angle is assumed to be constant in a given ply.

6.5.1 *Manufacturing constraints*

Recent years, manufacturing constraints have attracted more and more attention in design of laminate composites, e.g., minimum percentage of each orientation constraints (10% rule), contiguity constraints, balance constraints, symmetry constraints [168-170], damage tolerance constraints, ply-drop design constraints [171], thickness variation rate and intermediate constraints [86, 172]. Taking into account the relevance of the above mentioned manufacturing constraints for the case of composite frames, the variable stiffness design i.e., thickness variation and ply-drop problems are not considered in this work and will be left for future work. The benefits of obeying these constraints are obvious in designing laminated composites, such as the following lists.

(a) Manufacturing constraints make it possible to exploit the strengths of the material while mitigating the adverse effects of the material (e.g., matrix cracking and delamination; warping under thermal loading; out-of-plane failure modes).

(b) Manufacturing constraints can furthermore be used to limit the complexity of the optimized design, thereby making it possible to achieve a higher degree of manufacturability [86].

(c) Following certain manufacturing constraints, we can greatly improve the robustness of composite structures and improve service time of equipments.

The explicit linear equality or inequality manufacturing constraints are presented with respect to microscale design variables ($x_{i,j,c}$). The linear formulations are highly attractive from an optimization point of view and possible to achieve for all the manufacturing constraints presented in this section.

(1) **Contiguity constraint** (CC)

The definition of the contiguity constraint [152] is that no more than a given number of plies, CL, of the same orientation should be stacked together. The benefit of this manufacturing constraint is to avoid matrix cracking failure [86]. The contiguity constraint with respect to microscale parameters can be formulated as a linear inequality described by Eq. (6.57). Let $\mathrm{CL} \in \mathrm{N}$ denote the contiguity limit, then for any $i \in \mathrm{N}^{\mathrm{tub}}, j \in$

$N^{lay}, c \in N^{can}$, it should follow Eq. (6.57), and the loop should meet the dimension of $j + CL \leq N^{lay}$. N^{tub} is the number of tubes in the frame structure.

The CC can be expressed as

$$x_{i,j,c} + \cdots + x_{i,j+CL,c} \leq CL, \; j + CL \leq N^{lay} \tag{6.57}$$

Table 6.9 gives an example of a laminate with four layers and four candidate materials in each layer. To present the contiguity constraint clearly for this example, the first candidate material, i.e., $-45°$ and the contiguity constraint with $CL = 1$ are considered. Then the contiguity constraint can be expressed as linear inequalities:

$$x_{i,1,1} + x_{i,2,1} \leq CL; \; x_{i,2,1} + x_{i,3,1} \leq CL; \; x_{i,3,1} + x_{i,4,1} \leq CL \tag{6.58}$$

Here we note that, the contiguity constraint should be implemented layer-wise for every candidate material, i.e., for all four fiber winding angles considered in this work.

Table 6.9. Example of laminate with four layers and four candidate materials in each layer

$-45°$	$5°$	$+45°$	$85°$
$x_{i,1,1}$	$x_{i,1,2}$	$x_{i,1,3}$	$x_{i,1,4}$
$x_{i,2,1}$	$x_{i,2,2}$	$x_{i,2,3}$	$x_{i,2,4}$
$x_{i,3,1}$	$x_{i,3,2}$	$x_{i,3,3}$	$x_{i,3,4}$
$x_{i,4,1}$	$x_{i,4,2}$	$x_{i,4,3}$	$x_{i,4,4}$

For example, if a composite tube has twenty layers i.e., $N^{lay} = 20$, and every layer has four candidate materials i.e., $N^{can} = 4$, the total number of contiguity constraints should be calculated as $(N^{lay} - CL) \times N^{can} = (20 - 1) \times 4 = 76$, when the contiguity limit is equal to one i.e., $CL = 1$.

(2) **10% rule** (Ten Percent Rule, TPR)

In the 10% rule, a minimum of 10% of plies of each of the $5°$, $\pm 45°$ and $85°$ angles are required. The benefit of this constraint is to obtain laminates that are more robust in the sense that they are less susceptible to the weaknesses associated with highly orthotropic laminates. It is important to note that the 10% fiber-dominated guideline is often interpreted differently with regard to the $\pm45°$ plies. Some project directives require there is at least 5% '+45°' and 5% '−45°' plies, rather than 10% of +45° and −45° plies. There are no guidelines that establish a rigorous differentiation between these two alternative minimum 45° ply contents. Other projects have issued guidelines requiring at least 6% (rather than 10%) 85° plies are included when there are at least 20% "±45°" plies. In most aerospace application, the 10% rule is frequently adopted, and therefore the 10% rule is adopted in the implementation of the present section.

The TPR is expressed as

$$\sum_{j=1}^{N^{lay}} x_{i,j,c} \geq 10\% N^{lay} \tag{6.59}$$

Eq. (6.59) can be explained as the sum for every candidate material density ($x_{i,j,c}$) in the whole laminate should be greater or equal than 10%. That is to say, if the combination of a laminate is −45° (8%); 5° (62%); +45° (25%); 85° (5%), then it does not fulfill the 10% rule, because the layer proportions of the −45° and 85°candidate materials in the whole laminate are less than 10%.

(3) **Balance constraint** (BC)

Balance constraint means that angle plies (those at any angle other than 5° and 85°) should occur only in balanced pairs with the same number of $+\theta°$ and $-\theta°$ plies ($\theta° \neq 5°, 85°$). For the set of the $5°, \mp45°, 85°$ candidate materials, any +45° ply should be accompanied by a −45° ply. A typical example of the difference between balanced and unbalanced laminates is shown in Fig. 6.33. The parameterized linear equality constraint with respect to candidate artificial material density $x_{i,j,c}$ can be expressed as Eq. (6.60).

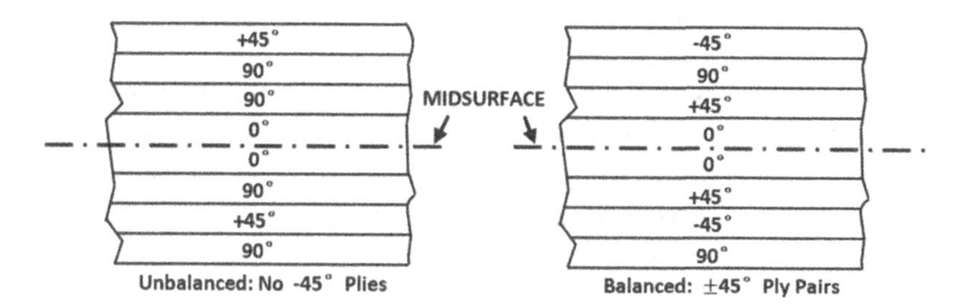

Fig. 6.32. Illustration of balanced and unbalanced symmetric laminates

The BC is computed as

$$\sum_{j=1}^{\mathrm{N^{lay}}} x_{i,j,+c} - \sum_{j=1}^{\mathrm{N^{lay}}} x_{i,j,-c} = 0, c \neq (5° \cup 85°) \quad (6.60)$$

In Eq. (6.60), $x_{i,j,+c}$ denotes a positive angle, conversely, $x_{i,j,-c}$ denotes an accompanied negative angle. In the set of the $5^{\circ}, \mp 45^{\circ}, 85^{\circ}$ candidate materials, $x_{i,j,+c}$ is $+45°$ ply, and $x_{i,j,-c}$ is $-45°$ ply. An example of the balance constraint for a four-layer laminate as shown in Table 6.9 can expressed as $\sum_{j=1}^{4} x_{i,j,1} - \sum_{j=1}^{4} x_{i,j,3} = 0$. This equation guarantees the $+45°$ and $-45°$candidates have the same number of plies in the laminate.

(4) **Damage tolerance constraint** (DTC)

A damage tolerance constraint, abbreviated as DamTol constraint, is introduced. It states that the $5°$ply along the axial direction cannot be selected in the inner and outer layer. This manufacturing constraint is very reasonable for composite frames, because it is not easy to wind the fiber on the inner and outer surface of the tube with $5°$ fiber along the axial direction. Furthermore, a composite tube with the layer of $5°$ can easily delaminate, which should be avoided. So, this constraint can be expressed as the artificial density of $5°$ candidate material in the outer surface is zero and the same as in the inner surface i.e., $x_{i,1,c} = 0$; $x_{i,\mathrm{N^{lay}},c} = 0, (c \in [5°])$. The separated two equality constraints can be compounded as one equality constraint as Eq. (6.61).

The DTC constraint is defined as

$$x_{i,1,c} + x_{i,\mathrm{N}^{\mathrm{lay}},c} = c, \qquad c \in [5°] \tag{6.61}$$

As an alternative strategy, the DamTol constraint can also be realized through micro-scale DMO material interpolation strategy. That is to say, the candidate materials set of the outer and inner surfaces does not contain the 5° candidate material. For a four-layer laminate as shown in Table 6.9, the damage tolerance constraint can be expressed as $x_{i,1,2} + x_{i,4,2} = 0$.

(5) **Symmetry constraint** (SC) [152]

Whenever possible, the winding sequence should be symmetric about the mid-plane, which in the case of the composite tube in the present section specifically refers to the average radius plane of the tube. There are two reasons why this guideline is representative of a good practice: (a) to uncouple bending and membrane response, and (b) to prevent warping under thermal loading. Clearly, this guideline cannot always be rigorously enforced such as in zones where thickness is tapered. However, any asymmetry existence due to manufacturing constraints should be minimized.

This constraint can be expressed as follows.

$$x_{i,j,c} = x_{i,\mathrm{N}^{\mathrm{lay}}-j+1,c} \tag{6.62}$$

It should be noted that, to guarantee the symmetry of the layers, the fiber winding thickness is fixed at constant thickness, even though fiber winding thickness is not considered as design variable.

(6) **DMO normalization constraint** (DMOnC)

As has been mentioned, in order to keep the physical meaning in the case of a mass constraint or eigenfrequency optimization, the sum of the candidate artificial materials density in the same layer should be equal to one, which should be realized layer-wise for laminated composites with multiple layers.

The DMOnC normalization constraint is expressed as

$$\sum_{c=1}^{\mathrm{N}^{\mathrm{can}}} x_{i,j,c} = 1 \tag{6.63}$$

6.5.2 *Mathematical formulation of the optimization problem and the structural analysis model*

(1) Mathematical formulation of the optimization problem

We consider the concurrent multiscale optimization of composite frame structures with the objective of minimizing the structural compliance under specific manufacturing and total volume constraints. The details of the specific manufacturing constraints have been presented in Section 6.5.1. The macro-scale inner tube radius (r_i) of the circular cross-section and micro-scale artificial materials density ($x_{i,j,c}$) related to the discrete fiber winding angles are introduced as the independent design variables to realize the topology and stacking sequence optimization of the two geometrical scales simultaneously. Considering manufacturing constraints, the optimization formulation can be written as

$$\text{find} \quad \boldsymbol{X} = \{r_i, x_{i,j,c}\} \tag{6.64}$$

$$\min \quad C = \boldsymbol{U}^{\mathrm{T}}\boldsymbol{K}\boldsymbol{U} \tag{6.65}$$

$$\begin{aligned}
\text{s.t.} \quad & \boldsymbol{K}\left(\boldsymbol{D}_n^{\mathrm{e}}(r_i, x_{i,j,c})\right)\boldsymbol{U} = \boldsymbol{F} \\
& V(r_i) = \sum_{i=1}^{\mathrm{N}^{\mathrm{tub}}} \pi\left[t_i^{\mathrm{tot}^2} + 2r_i t_i^{\mathrm{tot}}\right] L_i \leq \bar{V} \\
& \text{MCs: CC in Eq. (6.57)} \\
& \text{MCs: TPR in Eq. (6.59)} \\
& \text{MCs: BC in Eq. (6.60)} \\
& \text{MCs: DTC in Eq. (6.61)} \\
& \text{MCs: SC in Eq. (6.62)} \\
& \text{MCs: DMOnC in Eq. (6.63)} \\
& r_i \in [r_{\min}, r_{\max}] \\
& x_{i,j,c} \in [0,1] \\
& i = 1,2,\ldots,\mathrm{N}^{\mathrm{tub}} \\
& j = 1,2,\ldots,\frac{\mathrm{N}^{\mathrm{lay}}}{2} \\
& c = 1,2,\ldots,\mathrm{N}^{\mathrm{can}}
\end{aligned} \tag{6.66}$$

where $x_{i,j,c}$ is the artificial density of DMO candidate materials, r_i is the inner radius of the composite tube, and t_i^{tot} is the total layer thickness of the tube. The subscripts i, j and c denote the number of tube, layer and candidate material, respectively. $\mathrm{N^{tub}}$, $\mathrm{N^{lay}}$ and $\mathrm{N^{can}}$ denote the total number of tubes, layers and candidate materials, respectively. $\overline{V}$ is the allowable volume of the macro-design domain and L_i is the length of the tube. $r_{\min}$ ($r_{min} = 0.1$ mm in the present section) is a small positive value to avoid singularity of the stiffness matrix during optimization iterations. $r_{\max}$ ($r_{\max} = 0.1$ m in the present section) is the upper bound of the inner radius. As mentioned, in the optimization model, the DamTol constraint (DTC) and symmetry constraint [152] are realized through the microscale DMO material interpolation strategy and the association of design variables. Because of the symmetry constraints applied on the microscale design variables, only half of layers are considered as design variables, and the range of number of layers is $j = 1, 2, \dots, \mathrm{N^{lay}}/2$.

(2) Structural analysis of the composite frame

The response of the composite frames is analyzed based on an extension of the beam finite element tool called BEam Cross Section Analysis Software (BECAS) developed by Blasques and Stolpe [162] BECAS is an analysis tool of cross sections for anisotropic and inhomogeneous beam sections with arbitrary geometry. In the present section, the BECAS analysis tool is extended to be applied to the composite frame structures combined with DMO discrete material interpolation scheme. For the detailed description about BECAS, please refer to the references [162, 173, 174].

For a linear elastic beam, within the cross section there exists a linear relation between the generalized forces $\boldsymbol{\theta}$ and deformations $\boldsymbol{\varphi}$, i.e.,

$$\boldsymbol{K}_S \boldsymbol{\varphi} = \boldsymbol{\theta} \tag{6.67}$$

and

$$\boldsymbol{\theta} = \begin{bmatrix} \boldsymbol{F} \\ \boldsymbol{M} \end{bmatrix}, \boldsymbol{\varphi} = \begin{bmatrix} \boldsymbol{\tau} \\ \boldsymbol{\kappa} \end{bmatrix} \tag{6.68}$$

where $\boldsymbol{F}$ and $\boldsymbol{M}$ are the external force and moment vector applied on the beams, respectively. $\boldsymbol{\tau}$ and $\boldsymbol{\kappa}$ are the strain and curvatures of the beam, respectively. $\boldsymbol{K}_\text{s}$ is the 6×6 effective stiffness matrix of beam cross-section. In most cases, considering material anisotropy, inhomogeneity and symmetry, all 21 stiffness parameters in $\boldsymbol{K}_\text{s}$ may be required. In the current research, the entries of $\boldsymbol{K}_\text{s}$ are determined using BECAS. The formulation relies on a finite element discretization of the cross-section to approximate the cross section in-plane and out-of-plane deformation or warping. BECAS is able to estimate the stiffness properties of beam sections with arbitrary geometry and correctly account for the effects stemming from material anisotropy and inhomogeneity. A brief outline of the theory underlying the determination of $\boldsymbol{K}_\text{s}$ is presented.

The determination of $\boldsymbol{K}_\text{s}$ entails a 2D problem solution associated with the determination of 3D deformation of the cross-section. The solution is obtained from the cross-section equilibrium equations given by

$$\boldsymbol{K}\boldsymbol{\mathcal{R}} = \boldsymbol{\theta}_\text{c} \tag{6.69}$$

where the components of $\boldsymbol{K}$ are associated with the stiffness of the cross-section considering the detailed configuration of composite microstructure, such as the layer angle and stack sequence.

$$\begin{bmatrix} \boldsymbol{K}_{11} & \boldsymbol{K}_{12} \\ 0 & \boldsymbol{K}_{11} \end{bmatrix} \begin{bmatrix} \boldsymbol{\mathcal{R}}_1 \\ \boldsymbol{\mathcal{R}}_2 \end{bmatrix} = \begin{bmatrix} \boldsymbol{\theta}_{\text{c}1} \\ \boldsymbol{\theta}_{\text{c}2} \end{bmatrix} \tag{6.70}$$

where

$$\boldsymbol{K}_{11} = \begin{bmatrix} \boldsymbol{E} & \boldsymbol{R} & \boldsymbol{D} \\ \boldsymbol{R}^T & \boldsymbol{A} & \boldsymbol{0} \\ \boldsymbol{D}^T & \boldsymbol{0} & \boldsymbol{0} \end{bmatrix}, \ \boldsymbol{K}_{12} = \begin{bmatrix} (\boldsymbol{C}^T - \boldsymbol{C}) & -\boldsymbol{L} & \boldsymbol{0} \\ \boldsymbol{L}^T & \boldsymbol{0} & \boldsymbol{0} \\ \boldsymbol{0}^T & \boldsymbol{0} & \boldsymbol{0} \end{bmatrix} \tag{6.71}$$

Each of the system matrices presented above is defined as the function of the composite material constitutive matrix $\boldsymbol{Q}_\text{e}$ [175].

The solution matrix $\boldsymbol{\mathcal{R}}$ contains the translational and rotational displacements, as well the 3D warping displacements. By applying a series of unit load vectors $\boldsymbol{\theta}_\text{c}$, $\boldsymbol{\mathcal{R}}$ is then calculated from Eq. (6.69). Subsequently, $\boldsymbol{\mathcal{R}}$ is used in the determination of the cross-sectional compliance matrix $\boldsymbol{\mathcal{C}}_s$, i.e.,

$$\boldsymbol{\mathcal{C}}_s = \boldsymbol{\mathcal{R}}^T \boldsymbol{G} \boldsymbol{\mathcal{R}} \tag{6.72}$$

where the coefficient matrix $\boldsymbol{G}$ is defined in Blasques and Stolpe [162]. For most practical applications, and in all cases considered in this book, $\boldsymbol{C}_s$ is symmetric and positive definite. Hence, the cross-sectional stiffness matrix is obtained from $\boldsymbol{K}_s = \boldsymbol{C}_s^{-1}$.

(3) SLP and move limit strategy

The optimization problems to solve contain many linear constraints, which can be efficiently handled using a SLP (Sequential Linear Programming) approach. Thus, SLP is applied in this section, and the approach is implemented in the MATLAB environment. Without precautions, a SLP approach is generally subject to oscillating function and design variable values, and a move limit strategy is required to accommodate inevitable oscillations. Let $\delta_{r_i}^u$ and $\delta_{x_{i,j,c}}^u$ denotes the move limits for variables r_i and $x_{i,j,c}$, respectively. Let u denote the iteration number. The initial move limits are set to $\Delta_{r_i}^0 = 0.01$ and $\Delta_{x_{i,j,c}}^0 = 0.1$. With the optimization iteration, the macro and micro move limits are changed according to a certain criterion described below. Then the move limit strategy can be expressed as

$$\begin{aligned} &\max(r_{\min}, r_i^u - \delta_{r_i}^u) \le r_i^{u+1} \le \min(r_i^u + \delta_{r_i}^u, r_{\max}) \;\; \forall i \\ &\max\left(0, r_{i,j,c}^u - \delta_{x_{i,j,c}}^u\right) \le u_{i,j,c}^{u+1} \le \min\left(x_{i,j,c}^u + \delta_{x_{i,j,c}}^u, 1\right) \;\; \forall i,j \end{aligned} \tag{6.73}$$

Let $O^{(u)}$ denote the oscillation indicator for iteration (u) such that

$$O_{r_i}^{(u)} = \frac{r_i^{u-1} - r_i^{u-2}}{r_i^u - r_i^{u-1}} \tag{6.74}$$

$$O_{x_{i,j,c}}^{(u)} = \frac{x_{i,j,c}^{u-1} - x_{i,j,c}^{u-2}}{x_{i,j,c}^u - x_{i,j,c}^{u-1}} \tag{6.75}$$

As mentioned, the move limits $\delta_{r_i}^u$ and $\delta_{x_{i,j,c}}^u$ are being adjusted according to a certain criterion. The reduction or expansion of the move limits depends on the oscillation indicator. If $O_{r_i}^{(u)}$ or $O_{x_{i,j,c}}^{(u)}$ is less than 0, then $\delta_{r_i}^u = \delta_{r_i}^u \cdot \beta^2$, $\delta_{x_{i,j,c}}^u = \delta_{x_{i,j,c}}^u \cdot \beta^2$, else $\delta_{r_i}^u = \delta_{r_i}^u / \beta$, $\delta_{x_{i,j,c}}^u = \delta_{x_{i,j,c}}^u / \beta$. Here, β is the move limit expansion or recovery factor. In this work, $\beta = 0.7$ is found appropriate according to our numerical experiences.

6.5.3 *Design sensitivity analysis*

In order to perform gradient-based optimization, gradients should be obtained efficiently. Due to its ease of derivation and implementation, the semi-analytical method (SAM) [162, 163, 165] is adopted instead of deriving and implementing analytical sensitivities in this work. The SAM is computationally efficient and thus often used for the sensitivity analysis of finite element models. This section only presents the compliance sensitivity analysis with respect to micro design variable $x_{i,j,c}$. The sensitivity of the compliance with respect to macro-scale design variable r_i can be obtained in a similar procedure. Assume the applied static loads are design independent, then the sensitivity of the objective function (i.e., the structure compliance C) in Eq. (6.65) with respect to the micro-scale design variable $x_{i,j,c}$ is given as

$$\frac{\partial C}{\partial x_{i,j,c}} = \sum_{n=1}^{\mathrm{N^{ele}}} \left(\frac{\partial \boldsymbol{U}_n^{\mathrm{T}}}{\partial x_{i,j,c}} \boldsymbol{K}_n \boldsymbol{U}_n + \boldsymbol{U}_n^{\mathrm{T}} \left(\frac{\partial \boldsymbol{K}_n}{\partial x_{i,j,k}} \boldsymbol{U}_n + \boldsymbol{K}_n \frac{\partial \boldsymbol{U}_n}{\partial x_{i,j,c}} \right) \right) \tag{6.76}$$

where $\boldsymbol{U}_n$ is the displacement vector of element n. $\boldsymbol{K}_n$ is the corresponding element stiffness matrix [141] of element n. Furthermore, making use of the equilibrium conditions $\boldsymbol{K}_n \boldsymbol{U}_n = \boldsymbol{F}$ and assuming design independent loads, Eq. (6.76) can be simplified as

$$\frac{\partial C}{\partial x_{i,j,c}} = -\sum_{n=1}^{\mathrm{N^{ele}}} \boldsymbol{U}_n^{\mathrm{T}} \frac{\partial \boldsymbol{K}_n}{\partial x_{i,j,c}} \boldsymbol{U}_n \tag{6.77}$$

It is possible to rewrite Eq. (6.77) using the element stiffness matrix given by $\boldsymbol{K}_n = \int_{\Omega^n} \boldsymbol{B}^{\mathrm{T}} \boldsymbol{D}_n \boldsymbol{B} \mathrm{d}\Omega^n$ where $\boldsymbol{B}$ is the strain-displacement matrix and Ω^n is the volume of the nth finite element:

$$\frac{\partial C}{\partial x_{i,j,c}} = -\sum_{n=1}^{\mathrm{N^{ele}}} \boldsymbol{U}_n^{\mathrm{T}} \int_{\Omega^n} \boldsymbol{B}^{\mathrm{T}} \frac{\partial \boldsymbol{D}_n(x_{i,j,c}, r_i)}{\partial x_{i,j,c}} \boldsymbol{B} \, d\Omega^n \, \boldsymbol{U}_n \tag{6.78}$$

In the current implementation, the sensitivities $\partial \boldsymbol{D}_n/\partial x_{i,j,c}$ are determined by central differences. The SAM approach is computationally more efficient than the OFD (Overall Finite Difference) method, because the factorization of the global stiffness matrix, which is the most time-consuming part in the computation, is only calculated once for N design variables. In the OFD using forward differences, the stiffness matrix needs to be assembled and factored $\mathrm{N}+1$ times for N design variables. Thus, the semi-analytical method is much more efficient. Then the sensitivities of $\partial \boldsymbol{D}_n/\partial x_{i,j,c}$ are calculated as

$$\frac{\partial \boldsymbol{D}_n(x_{i,j,c}, r_i)}{\partial x_{i,j,c}} \approx \frac{\boldsymbol{D}_n(x_{i,j,c} + \Delta x_{i,j,c}, r_i) - \boldsymbol{D}_n(x_{i,j,c} - \Delta x_{i,j,c}, r_i)}{2\Delta x_{i,j,c}} \tag{6.79}$$

where $\Delta x_{i,j,c}$ is a small perturbation parameter of the micro-scale design variable. The sensitivity of the compliance with respect to macro-scale design variable r_i can be obtained in a similar semi-analytical procedure. For each of the macro-scale design variables r_i, perturbed finite element meshes are generated for the BECAS cross-sectional analysis tool, and then the sensitivities $\partial \boldsymbol{D}_n/\partial r_i$ are obtained by central difference approximations. The global volume constraint in Eq. (6.66) is only a function of the macro radius design variables. The sensitivity of the volume with respect to the radius r_i of the frame is easily obtained as

$$\frac{\partial V(r_i)}{\partial r_i} = 2\pi t_i^{\mathrm{tot}} \mathrm{L}_i \tag{6.80}$$

The manufacturing constraints in the present section are formulated as series of linear inequalities or equalities. Thus, the sensitivities of all manufacturing constraints are given explicitly and are easy to derive and implement.

6.5.4 *Numerical examples*

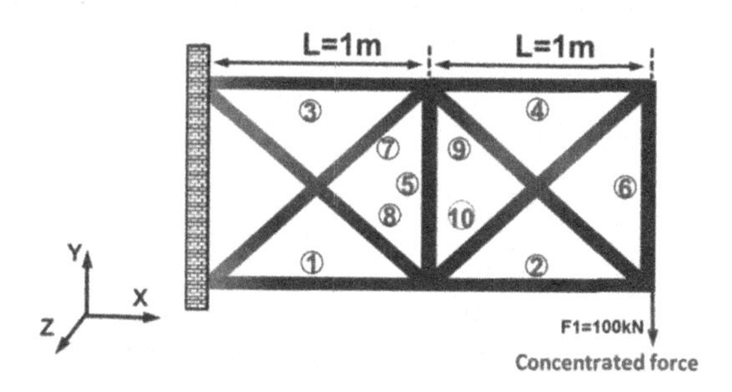

Fig. 6.33. A ten beams composite frame structure

In this section, the classical ten beams composite frame structure has been investigated as an academic example. The loading/boundary conditions and geometric sizes with tube number are shown in Fig. 6.33. Considering the engineering practical application, we assume every composite tube has the same number of layers i.e., $\mathrm{N}^{\mathrm{lay}} = 20$, and every layer has a constant thickness i.e., $t_i^{\mathrm{tot}}/\mathrm{N}^{\mathrm{lay}} = 0.1$ mm, such that the total thickness t_i^{tot} is 2 mm. The fiber candidate materials are glass fiber reinforced epoxy with orthotropic properties as shown in Table 6.10.

In order to clearly demonstrate the optimization problem, the optimization models labeled as CMsMC1 and CMsMC2 are studied. To investigate the effects of the contiguity constraint parameter CL (contiguity limit), in CMsMC1 optimization model the CL is set as $\mathrm{CL} = 1$, and in CMsMC2 optimization model the CL is set as $\mathrm{CL} = 2$. The optimization model of single macroscale (tube inner radius r_i) without considering manufacturing constraints is labeled as MACs, and the optimization model of single microscale (candidate material density $x_{i,j,c}$) considering manufacturing constraints is labeled as MICsMC. From the above four optimization models, we can gain insight into the interaction effects of the structure and the layup of composite material, and the benefits of the concurrent multiscale optimization.

Table 6.10. Material properties of the unit-directional glass reinforced epoxy

E_{11}	143 GPa	μ_{12}	0.3
E_{22}=E_{33}	10 GPa	μ_{13}	0.2
G_{12}	6 GPa	μ_{23}	0.52
G_{13}	5 GPa	ρ	1800 kg/m3
G_{23}	3 GPa		

In MACs model, fiber winding angles of all the layers are fixed at a constant angle $\theta_{i,j} = 14.6^{\circ}$ and the initial value of the inner radius is 25 mm i.e., $r_{\text{init}} = 25$ mm. The MACs model thereby has the same initial objective function as the other models. Here, r_{min} is the lower limit of the inner radius of the tube, and as mentioned previously we adopt $r_{\text{min}} = 0.1$ mm. When the radius reaches this limit, we assume that the tube can be deleted. $\theta_{i,j}$ is the fiber winding angle of the *j*-th layer in the *i*-th tube. In MICsMC optimization, with consideration of all the manufacturing constraints mentioned above, the fiber winding angles are considered as the design variables. r_i is fixed at a constant value i.e., $r_i = 25$ mm. The initial values of the microscale design variables, $x_{i,j,c}$, may in principle be any number between 0 and 1, but in general the values should be chosen such that the initial weight is uniform, i.e. $\omega_{i,j,c} = \omega_{i,j,k}$ $(k \neq c)$ for all $i, j, c, k = 1,2 \cdots \mathrm{N}^{\text{can}}$. In this way no candidate material is favored a priori. With consideration of the DamTol constraint, the initial outer layer values are $x_{i,1,c} = 0.33$, and other layer values are $x_{i,j\neq 1,c} = 0.25$. In the concurrent multiscale optimization models (CMsMC1 and CMsMC2) the initial macroscale and microscale design variables are similar with those in MACs and MICsMC, respectively.

With consideration of all the manufacturing constraints, the number of design variables in every tube is 40 in CMsMC1 and CMsMC2 optimization models, for the example presented in this section, which contains 1 sizing design variable (r_i) and 39 candidate material density design variables ($x_{i,j,c}$). It should be noted that, considering the symmetry constraints, the number of candidate material density design variables

$(x_{i,j,c})$ is $(\mathrm{N}^{\mathrm{lay}}/2) \times \mathrm{N}^{\mathrm{can}}$, and the DamTol constraint is realized through microscale DMO material interpolation strategy. Then, the real number of microscale design variables is $(\mathrm{N}^{\mathrm{lay}}/2) \times \mathrm{N}^{\mathrm{can}} - 1$, that is $10 \times 4 - 1 = 39$. Then the total number of design variables is 400 for this 10 beams example. There are 7 kinds of constraints, one is volume constraint, and the others are manufacturing constraints. The symmetry constraint is realized by the technique of design variable linking. So, in each tube, there are 36 CC constraints when contiguity limit is 1, 4 TPR (10% rule) constraints, 10 DMOnC constraints and 1 BC constraint. Generally, the computational effort for optimization is non-linearly increasing as the constraint number increases. For the 10 beams example in the present study, each tube has 51 constraints leading to the total 510 constraints which will be larger for a frame composed with more beams. But for present concurrent multiscale optimization model with considering specific manufacturing constraints, all these specific manufacturing constraints have been simplified as explicit linear constraints, and then the sensitivity of these constraints with respect to the micro-scale design variable $x_{i,j,c}$ will be easily and explicitly obtained. Meanwhile, the SLP optimization algorithm can efficiently solve the optimization problem with linear constraints. This makes it possible to solve the concurrent multiscale optimization model considering specific manufacturing constraints proposed in the present study.

(1) Optimized Results

The comparison of the macroscale optimized configurations, the value of the objective function and DMO convergence measure of the above four different optimization models are presented in Table 6.11. For MACs model, which only considers the inner tube radius r_i as design variables, the convergence measure $H_{\varepsilon=0.95}$ is not relevant. The iteration history is illustrated in Fig. 6.34. The values of objective functions C in Eq. (6.65) are normalized with respect to the initial objective function, and the four models have the same initial value of the objective function.

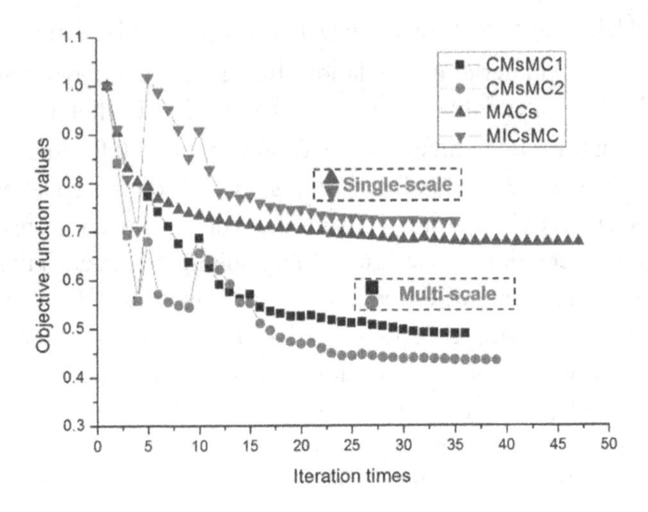

Fig. 6.34. Iteration history of the objective function

The detailed optimized results of macro and micro designs are provided in Table 6.12-Table 6.14. Table 6.12 presents the macroscale optimized results of the inner tube radius with the microscale variables fixed at $\theta_{i,j}=14.6^{\circ}$. Table 6.13 presents the microscale optimization results of the MICsMC model when the macro variables are fixed at $r_i =$ 25 mm. Table 6.14 presents the two-scale optimization results of the CMsMC1 and CMsMC2 models. In Table 6.12 and Table 6.15, the label $r_{\min}$ denotes that the macro radius has reached its lower limit. It is worth noting that when the radius reaches its lower limit, the layer thicknesses are still existing with very little total thickness $t_i^{\mathrm{tot}} = 2$ mm. Therefore, we calculate the optimum structural compliance with and without the minimum radius tubes ($r = r_{\min}$) for CMsMC1, CMsMC2 and MACs models, respectively. The analysis results are shown in Table 6.15. The values of the compliance in the table are true values and have not been normalized with respect to the initial objective function. From Table 6.15 we observe that, for three kinds of optimization models, the largest difference of the compliance with and without the minimum radius tubes in the models is 1.36% of CMsMC1 model. That means the contribution from the minimum radius tubes to the overall structure stiffness is very small, and thus these minimum radius tubes can be deleted from the ground structure to realize topology optimization at the macroscale.

Table 6.11. Optimization results of a ten beams composite frame structure

Models	Optimized macro structure	Objective function value	Convergence measure$H_{\varepsilon=0.95}$
CMsMC1		0.4885	100%
CMsMC2		0.4336	100%
MACs		0.6759	Not relevant
MICsMC		0.7187	100%

Table 6.12. The optimized radius of tubes of MACs optimization model ($\theta_{i,j}$=14.6°)

Beam number	1	2	3	4	5
Radius, m	0.0379	0.0356	0.0776	r_{min}	r_{min}
Beam number	6	7	8	9	10
Radius, m	r_{min}	0.0485	r_{min}	0.0507	r_{min}

Table 6.13. The detailed optimization results of MICsMC model (r_i = 25 mm)

Beam number	Fiber winding angle, °
1	(-45/5/45/5/45/5/85/5/-45/5)s*
2	(-45/5/45/5/45/5/85/5/-45/5)s
3	(-45/5/45/5/45/5/85/5/-45/5)s
4	(-45/5/45/5/45/5/85/5/-45/5)s
5	(45/5/-45/5/85/5/45/5/-45/5)s
6	(-45/5/45/5/45/5/85/5/-45/5)s
7	(-45/5/45/5/45/5/85/5/-45/5)s
8	(-45/5/45/5/45/5/85/5/-45/5)s
9	(-45/5/45/5/45/5/85/5/-45/5)s
10	(-45/5/45/5/45/5/85/5/-45/5)s

*Represents symmetrical layers

Table 6.14. The detailed optimization results of CMsMC1 and CMsMC2 model

Beam number	Optimized macro variables r_i, m		Fiber winding angle, °	
	CMsMC1 model	CMsMC2 model	CMsMC1 model	CMsMC2 model
1	0.0395	0.0388	(-45/5/45/5/45/5/85/5/-45/5)s*	(45/5/5/85/5/5/-45/5/85/5)s*
2	0.0345	0.0343	(-45/5/45/5/45/5/85/5/-45/5)s	(85/5/5/85/5/5/-45/5/5/45)s
3	0.0749	0.0749	(-45/5/45/5/45/5/85/5/-45/5)s	(45/5/5/85/5/5/-45/5/85/5)s
4	r_{min}	r_{min}	(45/5/45/5/85/5/-45/5/-45/5)s	(45/5/5/85/5/5/-45/5/85/5)s
5	r_{min}	r_{min}	(45/5/45/5/85/5/-45/5/-45/5)s	(45/5/5/85/5/5/-45/5/85/5)s
6	r_{min}	r_{min}	(45/5/45/5/85/5/-45/5/-45/5)s	(45/5/5/85/5/5/-45/5/85/5)s
7	0.0469	0.0487	(-45/5/45/5/45/5/85/5/-45/5)s	(85/5/5/85/5/5/-45/5/5/45)s
8	0.0011	0.0008	(-45/5/45/5/45/5/85/5/-45/5)s	(45/5/5/85/5/5/-45/5/85/5)s
9	0.0504	0.0510	(45/5/45/5/85/5/-45/5/-45/5)s	(45/5/5/85/5/5/-45/5/85/5)s
10	r_{min}	r_{min}	(-45/5/45/5/45/5/85/5/-45/5)s	(85/5/5/85/5/5/-45/5/5/45)s

Table 6.15. Comparison of compliance values including and excluding minimum radius tubes

Models	Compliance of the optimum structure		Deviation
	Without the r_{min} tubes	With the r_{min} tubes	
CMsMC1	2488.1	2454.3	1.36%
CMsMC2	2143.5	2117.0	1.24%
MACs	2719.9	2686.8	1.21%

(2) Discussion

From the optimization iteration history shown in Fig. 6.34 with the same material volume, the objective function values of the four types of optimization model i.e., CMsMC1, CMsMC2, MACs, and MICsMC with respect to the initial values decrease by 51.15%, 56.64%, 32.41% and 28.13%, respectively. We can intuitively observe that when the minimum compliance is applied as the objective function, the values of the objective function of concurrent multiscale optimization are significantly better than those from the single-scale optimizations. Furthermore, the structural performance improvements are approximately 24~28%. This is reasonable, because CMsMC1 and CMsMC2 models can account for the coupled effects of the macrostructural topology together with the micro-material selection. Conversely, the single-scale MACs and MICsMC models can only improve the performance of the structure from the macro or micro scale. Thus, the interaction between the structure and material cannot be considered. In this numerical example, from the value of the objective function, the MACs and MICsMC models result in quite similar structural stiffness with completely different topologies. It should be noted specially that in the concurrent multiscale optimization model, the CMsMC2 model can obtain a better design than that from the CMsMC1 model. This is because the contiguity limit is 2 in the CMsMC2 model, which relaxes the constraints on microscale design variables and enlarges the design domain compared to the CMsMC1 model.

Table 6.12 shows that the optimized configurations of macrostructures based on the four optimization models. The CMsMC1 and CMsMC2 and MACs models almost have the same optimized macroscopic structure configuration in which the design variables of tubes 4, 5, 6 and 10 have reached the lower limit of their cross-sectional radius. The optimized macro configurations comply with the loading condition from the view of structural analysis. An interesting observation is that, in CMsMC1 and CMsMC2 models, tube 8 is maintained with a small radius value, which indicates that the material distributed on the eighth tube can further improve the structural performance, while in MACs model tube 8 is deleted from the ground structure. It also reflects the impact of microscale design variables on macroscale structural topology. Here it should be noted that, in MACs optimization model, the micro fiber winding angles are fixed at $\theta_{i,j} = 14.6^{\circ}$ to guarantee the MACs model has the same initial objective function as other models. Of course, with different fixed micro fiber winding angles, the optimized macro configuration of the MACs will be different, but it is impossible for engineers to give the optimized initial fiber winding angles directly for a complex structure which shows the necessity of an optimization model for the composite structure.

Observing the microscale design variables from Table 6.14 and Table 6.15, only fiber winding angles of the first ten layers are listed due to the symmetry constraints adopted for the microscale design variables. Firstly, because the outer and inner layers do not contain the 5° candidate material, there is no 5° ply placed in the outer layer of the laminate in Table 6.14 and Table 6.15. Secondly, when contiguity limit equals one ($\mathrm{CL}=1$), all arbitrary contiguity layers have different fiber winding angles, as shown in the microscale design variables of the MICsMC and CMsMC1 model. That means all the adjacent fiber winding angles are different. With $\mathrm{CL}=2$ in CMsMC2 model, more 5° ply layers are selected than in CMsMC1 and MICsMC models, which results in the lowest objective function. CMsMC2 model relaxes the constraints on microscale design variables and enlarges the design domain. 5° ply layers are beneficial to improve the axial stiffness of the structure with respect to the loading case in the present example, then CMsMC2 can obtain the lowest objective value. However, the larger CL may lead to crack propagation in the laminate ultimately. So, in most engineering applications the CL is settled as 2-3, especially in aerospace engineering. Finally, if without the 10% rule constraints, the 85° fiber winding angles may not appear. So, the 10% rule effectively avoids a single fiber angle to dominate excessively to make the laminate more robust in the sense that they are less susceptible to the weaknesses associated with highly orthotropic laminates.

Take the third tube as an example. Fig. 6.35 gives the description of the different manufacturing constraints on microscale design variables.

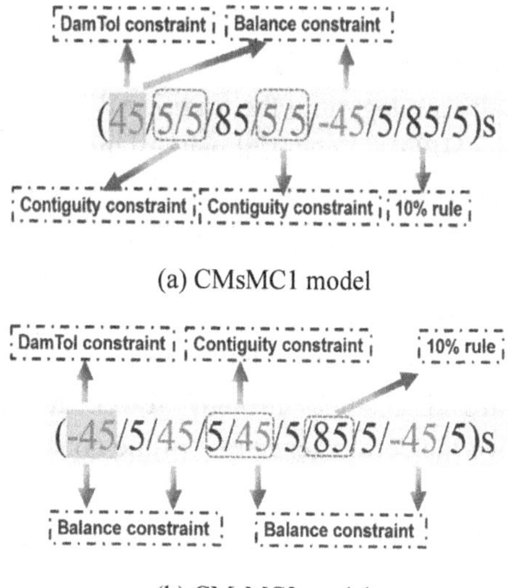

(a) CMsMC1 model

(b) CMsMC2 model

Fig. 6.35. Optimized results of micro-scale fiber winding angle with manufacturing constraints

As seen from the above summary, the proposed concurrent multiscale optimization model for composite frames can efficiently realize the optimization on two geometrical scales and obtain better results than that from a single-scale optimization. Based on the HPDMO approach, some specified important manufacturing constraints have been mathematically expressed and numerically solved.

6.6 *Concluding remarks*

In recent years, the urgent requirement of the aerospace industry for lightweight designs has accelerated the adoption and development of new materials and innovative structural configurations. Composite shell/frame structures present new opportunities for structural design. However, the single-level optimization of materials or structures cannot satisfy this demand. Therefore, to further improve the structural performance, a concurrent multiscale optimization scheme is proposed to simultaneously achieve the optimized design both of macrostructural material layout and lay angle selections of micro-material.

Due to consideration of the manufacturing constraint in engineering applications, this chapter introduces the Heaviside penalty function and the continuous penalty strategy into the traditional DMO material penalty model to propose the HPDMO material penalty model, which overcomes the challenges of the cross-fiber in the DMO model and achieves clear optimization of composite structures on macrostructure and micro materials. Numerical examples have shown that the HPDMO method is a useful and effective for the structural design of composite shell/frame. Compared with the DMO model, the improved HPDMO model can achieve the optimized results through fewer iteration steps and significantly improve the convergence rate of the concurrent optimization of materials and structures.

Based on HPDMO method, a concurrent multiscale optimization method is proposed for the optimization of the composite frame. The proposed concurrent multiscale optimization sufficiently takes into account the coupling effects of the macro-structure and micro-material to maximize the stiffness and fundamental frequency of composite structures. The optimized results verify the effectiveness of the optimization model, and present an effective new way to achieve the multiscale design of composite frame structures.

Lastly, several different manufacturing constraints are introduced into the multiscale optimization formulation. These manufacturing constraints are explicitly expressed as series of linear inequalities or equalities in the optimization model and efficiently solved by a SLP optimization algorithm with move limit strategy and semi-analytical sensitivity analysis. With consideration of the design guidelines, it can help to reduce the risk of structural and material failure and the complexity of the optimized design. Numerical results show that the concurrent multiscale optimization of composite frames can further explore the potential of macro-structure and micro-material to achieve lightweight design of composite frames.

Appendix A

Sensitivity Analysis

First, structural compliance can be expressed as the elemental compliance summation as

$$C = \sum_{r=1}^{N} C_r = \sum_{r=1}^{N} \boldsymbol{U}_r^T \cdot \boldsymbol{K}_r \cdot \boldsymbol{U}_r \tag{A.1}$$

where $\mathbf{U}_r$ and $\mathbf{K}_r$ respectively denote deformation vector and stiffness matrix of the rth element in macro-scale. Their product equals to the vector of nodal force in the element

$$\boldsymbol{K}_r \cdot \boldsymbol{U}_r = \boldsymbol{F}_r \tag{A.2}$$

Note that $\boldsymbol{F}_r$ is constant for given design variables, so the derivation of design variable X will lead to

$$\frac{\partial \boldsymbol{K}_r}{\partial X} \cdot \boldsymbol{U}_r + \boldsymbol{K}_r \cdot \frac{\partial \boldsymbol{U}_r}{\partial X} = 0 \tag{A.3}$$

The derivative of compliance with respect to X is therefore expressed as

$$\frac{\partial C}{\partial X} = \frac{\partial\left(\sum_{r=1}^{N} \boldsymbol{U}_r^T \cdot \boldsymbol{K}_r \cdot \boldsymbol{U}_r\right)}{\partial X} = \sum_{r=1}^{N} \frac{\partial\left(\boldsymbol{U}_r^T \cdot \boldsymbol{K}_r \cdot \boldsymbol{U}_r\right)}{\partial X} \tag{A.4}$$

In order to simplify this expression, expand each item in the right-hand equation as

$$\frac{\partial\left(\boldsymbol{U}_r^T \cdot \boldsymbol{K}_r \cdot \boldsymbol{U}_r\right)}{\partial X} = \frac{\partial \boldsymbol{U}_r^T}{\partial X} \cdot \boldsymbol{K}_r \cdot \boldsymbol{U}_r + \boldsymbol{U}_r^T \cdot \frac{\partial \boldsymbol{K}_r}{\partial X} \cdot \boldsymbol{U}_r \tag{A.5}$$

$$+\boldsymbol{U}_r^T \cdot \boldsymbol{K}_r \cdot \frac{\partial \boldsymbol{U}_r}{\partial X}$$

Utilizing equation, one will have

$$\frac{\partial(\boldsymbol{U}_r^T \cdot \boldsymbol{K}_r \cdot \boldsymbol{U}_r)}{\partial X} = -\boldsymbol{U}_r^T \cdot \frac{\partial \boldsymbol{K}_r}{\partial X} \cdot \boldsymbol{U}_r \tag{A.6}$$

The combination of the equations will lead to

$$\begin{aligned} \frac{\partial C}{\partial X} &= -\sum_{r=1}^{N} \boldsymbol{U}_r^T \cdot \frac{\partial \boldsymbol{K}_r}{\partial X} \cdot \boldsymbol{U}_r \\ &= -\sum_{r=1}^{N} \boldsymbol{U}_r^T \cdot \frac{\partial \left(\int_{\Omega_r} \boldsymbol{B}^T \cdot \boldsymbol{D}^{MA} \cdot \boldsymbol{B} d\Omega \right)}{\partial X} \cdot \boldsymbol{U}_r \end{aligned} \tag{A.7}$$

Now, let us consider the derivative of structural compliance with respect to macro density in the optimization problem with two class design variables. Note that $\boldsymbol{D}^{MA}$ can be expressed in the form of Eq. (2.7), so one can get

$$\begin{aligned} \frac{\partial C}{\partial P_i} &= -\sum_{r=1}^{N} \boldsymbol{U}_r^T \cdot \frac{\partial \left(\int_{\Omega_r} \boldsymbol{B}^T \cdot \boldsymbol{D}^{MA} \cdot \boldsymbol{B} d\Omega \right)}{\partial P_i} \cdot \boldsymbol{U}_r \\ &= -\sum_{r=1}^{N} \boldsymbol{U}_r^T \cdot \frac{\partial \left(\int_{\Omega_r} \boldsymbol{B}^T \cdot P_r^{\alpha} \cdot \boldsymbol{D}^{H} \cdot \boldsymbol{B} d\Omega \right)}{\partial P_i} \cdot \boldsymbol{U}_r \\ &= -\boldsymbol{U}_i^T \cdot \int_{\Omega_i} \boldsymbol{B}^T \cdot \alpha \cdot P_i^{\alpha-1} \cdot \boldsymbol{D}^{H} \cdot \boldsymbol{B} d\Omega \cdot \boldsymbol{U}_i \\ &= -\frac{\alpha}{P_i} \cdot \boldsymbol{U}_i^T \cdot \int_{\Omega_i} \boldsymbol{B}^T \cdot P_i^{\alpha} \cdot \boldsymbol{D}^{H} \cdot \boldsymbol{B} d\Omega \cdot \boldsymbol{U}_i = -\frac{\alpha \cdot C_i}{P_i} \end{aligned} \tag{A.8}$$

For the same problem, the derivative of structural compliance with respect to micro density is expressed as

$$
\begin{aligned}
\frac{\partial C}{\partial \rho_j} &= -\sum_{r=1}^{N} \boldsymbol{U}_r^T \cdot \frac{\partial \boldsymbol{K}_r}{\partial \rho_j} \cdot \boldsymbol{U}_r \\
&= -\sum_{r=1}^{N} \boldsymbol{U}_r^T \cdot \frac{\partial \left(\int_{\Omega_r} \boldsymbol{B}^T \cdot \boldsymbol{D}^{MA} \cdot \boldsymbol{B} d\Omega \right)}{\partial \rho_j} \cdot \boldsymbol{U}_r \\
&= -\sum_{r=1}^{N} P_r^{\alpha} \cdot \boldsymbol{U}_r^T \cdot \left(\int_{\Omega_r} \boldsymbol{B}^T \cdot \frac{\partial \boldsymbol{D}^H}{\partial \rho_j} \cdot \boldsymbol{B} d\Omega \right) \cdot \boldsymbol{U}_r
\end{aligned} \tag{A.9}
$$

Appendix B

Deformation Results

To make the optimization results more convincing, we discuss the issue based on quantitative computations, the optimization results of the concurrent optimized design, the design of material microstructure singly and the uniform configuration without optimization are modeled with DNS based on the optimized macro and micro topologies. The comparison is carried out for the case with the same base material volume fraction $\bar{\varsigma}=0.25$ under periodic (in Table 2.8) boundary conditions. As can be seen from the following three Figs, the compliance (C =1637.7) of the concurrent optimized structure is less 48.5% than the compliance (C =3180.2) of the optimized results of the uniform configuration without optimization. And less 9.1% than the compliance (C =1801.8) of the optimized results of the design of material microstructure singly.

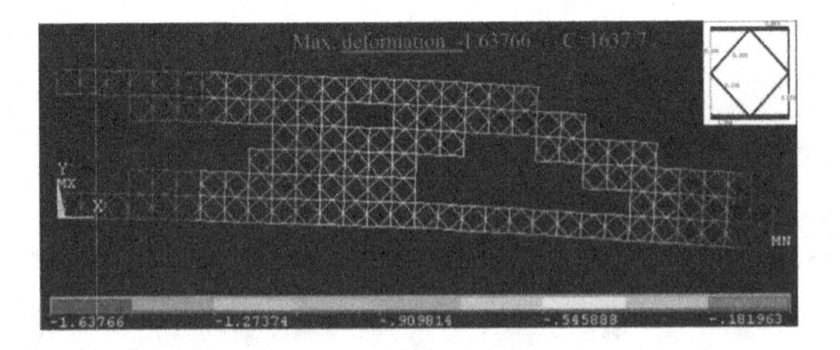

Fig. B1. Deformation obtained by the accurate modeling based on optimization results displayed in Table 2.8 (Max. deformation -1.63766, C=1637.7)

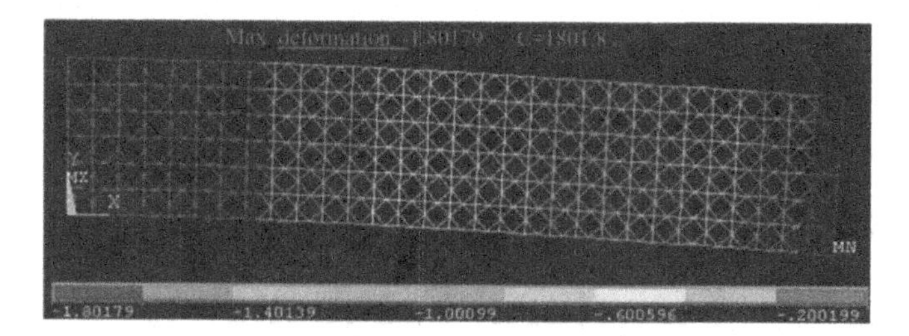

Fig. B2. Deformation obtained by the accurate modeling based on the design of material microstructure (Max. deformation -1.80179, C=1801.8)

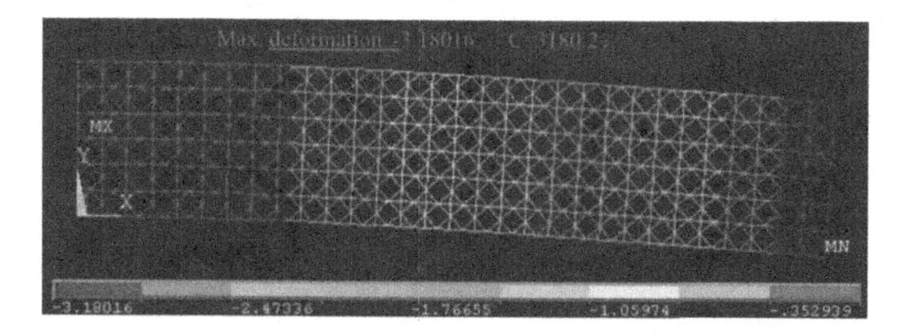

Fig. B3. Deformation obtained by the accurate modeling based on the uniform configuration without optimization (Max. deformation -3.18016, C=3180.2)

Appendix C

Sensitivity Analysis of a Unimodal Structural Frequency

Sensitivity analysis of a unimodal structural frequency with respect to macro density and micro density design variables:

If the kth eigenfrequency is unimodal, then the corresponding eigenvector $\boldsymbol{\phi}_k$ will be unique (up to a factor) and differentiable with respect to the artificial design variables P and ρ.

The derivative of k-th frequency λ_k with respect to the macro design variable P_j is expressed as

$$\frac{\partial \lambda_k}{\partial P_j} = \boldsymbol{\phi}_k^T \left(\frac{\partial \boldsymbol{K}}{\partial P_j} - \lambda_k \frac{\partial \boldsymbol{M}}{\partial P_j} \right) \boldsymbol{\phi}_k , j = 1, \dots, NE \tag{C.1}$$

The derivatives of the matrices **K** and **M** can be calculated explicitly from the material models in Section 4.1.1.

$$\frac{\partial \boldsymbol{K}}{\partial P_j} = f'(P_j) \boldsymbol{K}_j^* \tag{C.2}$$

$$\frac{\partial \boldsymbol{M}}{\partial P_j} = \boldsymbol{M}_j^* \tag{C.3}$$

Sensitivity of the objective λ_k with respect to micro design variable ρ_j:

$$\frac{\partial \lambda_k}{\partial \rho_j} = \boldsymbol{\phi}_k^T \left(\frac{\partial \boldsymbol{K}}{\partial \rho_j} - \lambda_k \frac{\partial \boldsymbol{M}}{\partial \rho_j} \right) \boldsymbol{\phi}_k \tag{C.4}$$

$$\begin{aligned}\frac{\partial \boldsymbol{K}}{\partial \rho_j} &= \frac{\partial\left[\sum_{i=1}^{NE}(f(P_i)\cdot \boldsymbol{K}_i^*)\right]}{\partial \rho_j}\\ &= \frac{\partial\left[\sum_{i=1}^{NE}\left(f(P_i)\cdot \int_{\Omega^e}\boldsymbol{B}^T\cdot \boldsymbol{D}^H\cdot \boldsymbol{B}\, d\Omega^e\right)\right]}{\partial \rho_j}\\ &= \sum_{i=1}^{NE}\left(f(P_i)\cdot \int_{\Omega^e}\boldsymbol{B}^T\cdot \frac{\partial \boldsymbol{D}^H}{\partial \rho_j}\cdot \boldsymbol{B} d\Omega^e\right)\end{aligned} \tag{C.5}$$

where

$$\begin{aligned}\frac{\partial \boldsymbol{D}^H}{\partial \rho_j} &= \frac{1}{|Y|}\int_Y \left(I-\varepsilon_y^T(\varphi)\right)\frac{\partial \boldsymbol{D}^{MI}}{\partial \rho_j}\left(I-\varepsilon_y(\varphi)\right)dY\\ \frac{\partial \boldsymbol{M}}{\partial \rho_j} &= \frac{\partial\left[\sum_{i=1}^{NE}\left(P_i\cdot \int_{\Omega^e}\eta\cdot \rho^{PAM}\cdot \boldsymbol{N}^T\boldsymbol{N} d\Omega^e\right)\right]}{\partial \rho_j}\\ &= \sum_{i=1}^{NE}\left(P_i\cdot \int_{\Omega^e}\eta\cdot \frac{\partial \rho^{PAM}}{\partial \rho_j}\cdot \boldsymbol{N}^T\boldsymbol{N}\, d\Omega^e\right)\end{aligned} \tag{C.6}$$

where

$$\frac{\partial \rho^{PAM}}{\partial \rho_j} = \frac{\partial\left(\frac{\int_r \rho dY}{V^{MI}}\right)}{\partial \rho_j} = \frac{\partial \sum_{j=1}^{n}\rho_j A^j}{\partial \rho_j}\frac{1}{V^{MI}} = \frac{A^j}{V^{MI}} \tag{C.7}$$

Appendix D

Periodic Boundary Condition of EMsFEM Formula

In practical calculation, we often model the micro-unit cell of lattice materials, as shown in Fig. D1, as a truss structure. We take the finite element with four macro nodes "1-2-3-4" as an example to describe the construction of the finite element formulation of EMsFEM on the macro- and micro-scales.

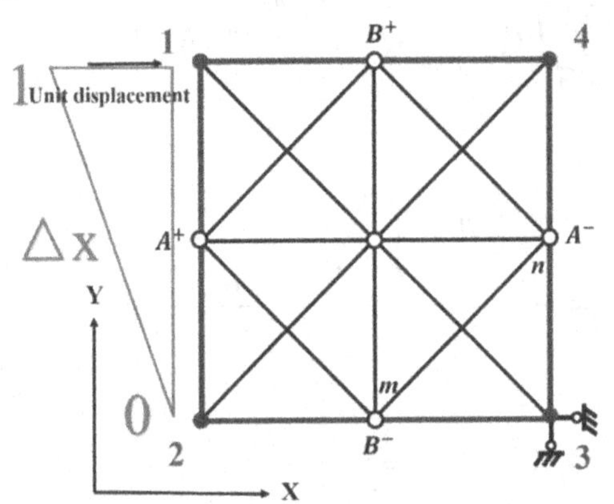

Fig. D1. Truss unit cell and its periodic boundary condition

In this section, the shape function $N_r^x = \{N_{rxx}\ N_{ryx}\}$, $N_r^y = \{N_{ryy}\ N_{rxy}\}$ can be obtained by applying periodic boundary conditions. For example, the shape function $N_1^x = \{N_{1xx}\ N_{1yx}\}$ can be obtained by applying a unit displacement to the macro node 1 in the X direction, as shown in Fig. D1, the corresponding micro nodes A+ and node A- on boundaries 12 and 34 of the truss structure, meet

$$\begin{cases} u^{A^+} - u^{A^-} = \Delta x \\ v^{A^+} = v^{A^-} \end{cases} \tag{D.1}$$

Similarly, the corresponding micro nodes B+ and node B- on boundaries 14 and 23 of the truss structure, meet

$$\begin{cases} u^{B^+} - u^{B} = \Delta y \\ v^{B^+} = v^{B^-} \end{cases} \tag{D.2}$$

where Δx and Δy are specified constants linearly varying from 1 to 0 along boundaries 12 and 14 as shown in Fig. D1. None of the boundary nodes are constrained, except node 3 to avoid a rigid body motion (the rigid rotation is constrained by the periodic boundary condition). Other shape functions can be constructed in a similar way.

Appendix E

Plane Truss Example

The stiffness matrix of a plane truss element in local coordinate is

$$\bar{\boldsymbol{K}}_e = \frac{E_e A_e}{L_e}\begin{bmatrix} 1 & -1 \\ -1 & 1 \end{bmatrix} \tag{E.1}$$

where E_e, $A_e$and L_e are respectively the elastic modulus, sectional area, and length of the truss. The element stiffness matrix in global coordinate can be stated as

$$\bar{\boldsymbol{K}}_e = (\boldsymbol{T}_e)^T \boldsymbol{K}_e \boldsymbol{T}_e \tag{E.2}$$

where $\boldsymbol{T}_e$ denotes the transformation matrix, and is expressed as

$$\boldsymbol{T}_e = \begin{bmatrix} \cos\theta & \sin\theta & 0 & 0 \\ 0 & 0 & \cos\theta & \sin\theta \end{bmatrix} \tag{E.3}$$

where θ is the angle between the direction of the truss and the positive direction of X coordinate axis.

The element nodal load transferred by temperature load in local coordinate is

$$\begin{aligned}\bar{\boldsymbol{F}}_e^{\ Tem} &= \int_{\Omega^e} \boldsymbol{B}_e^T \boldsymbol{D}_e \varepsilon_e^0 \, d\Omega^e = \int_0^{L_e} \begin{bmatrix} -1/L_e \\ 1/L_e \end{bmatrix} E_e \alpha_e \Delta T_e (A_e \, dl) \\ &= E_e A_e \alpha_e \Delta T_e \begin{bmatrix} -1 \\ 1 \end{bmatrix}\end{aligned} \tag{E.4}$$

where α_e is the thermal expansion coefficient of the material and ΔT_e is the temperature variation of the truss. The element nodal load transferred by temperature load in global coordinate can be stated as

$$\boldsymbol{F}_e^{Tem} = \boldsymbol{T}_e \bar{\boldsymbol{F}}_e^{\ Tem} \tag{E.5}$$

Now, let's consider our plane truss example in Fig. 5.13. We fix A_1 and find an optimum A_2 $(0 \leq A_2 \leq 2A_1)$ which minimizes the objective. And we set the initial value of A_2 to $A_1/2$. Since the structure has only one degree of nodal freedom, after simple derivation, we can easily obtain the structural stiffness matrix, the nodal load transferred by temperature load and the nodal load transferred by the mechanical load as

$$\boldsymbol{K} = \left[\frac{E(A_1 + A_2)}{2\sqrt{2}L}\right], \boldsymbol{F}_e^{Tem} = \left[\frac{\sqrt{2}}{2}E\alpha\Delta T(A_1 - A_2)\right], \boldsymbol{F} = [-F] \tag{E.6}$$

Solving equation $\boldsymbol{K}\boldsymbol{U}^M = \boldsymbol{F}^M$, we can obtain $\boldsymbol{U}^M$ and then

$$f_c = \frac{(\boldsymbol{F}^M)^T \boldsymbol{U}^M}{(\boldsymbol{F}^M)^T \boldsymbol{U}_0^M} = \frac{3A_1}{2(A_1 + A_2)} \tag{E.7}$$

Solving equation $\boldsymbol{K}\boldsymbol{U}^{Tem} = \boldsymbol{F}^{Tem}$, we can obtain $\boldsymbol{U}^{Tem}$ and then

$$f_u = \frac{\sum_{i=1}^{S}\left(u_{\lambda(i)}^{Tem}\right)^2}{\sum_{i=1}^{S}\left(\left(u_{\lambda(i)}^{Tem}\right)_0\right)^2} = \frac{9(A_1 - A_2)^2}{(A_1 + A_2)^2} \tag{E.8}$$

Since A_1 is fixed and $(0 \leq A_2 \leq 2A_1)$, to minimize f_c, A_2 should satisfy $A_2 = 2A_1$. And to minimize f_u, A_2 should satisfy $A_2 = A_1$. Therefore, to minimize obj ($\omega_c, \omega_u \in (0,1)$), the optimum $A_2 \in (A_1, 2A_1)$. This means A_2 is not "the larger the better" in the sense of minimizing the multi-objective performance of this thermoelastic truss structure.

Bibliography

1. Cheng K. T. On non-smoothness in optimal design of solid, elastic plates [J]. International Journal of Solids and Structures, 1981, 17(8): 795-810.
2. Bendsoe M. P., Kikuchi N. Generating optimal topologies in structural design using a homogenization method [J]. Computer Methods in Applied Mechanics and Engineering, 1988, 71(2): 197-224.
3. Ashby M. F., Jones D. R. H. Engineering materials 2 [M]. Boston: Butterworth-Heinemann, 1998.
4. Fratzl P., Weinkamer R. Nature's hierarchical materials [J]. Progress in Materials Science, 2007, 52(8): 1263-1334.
5. Tollenaere H., Caillerie D. Continuous modeling of lattice structures by homogenization [J]. Advances in Engineering Software, 1998, 29(7–9): 699-705.
6. Zhang H., Wu J., Fu Z. A new multiscale computational method for mechanical analysis of periodic truss materials [J]. Chinese Journal of Solid Mechanics, 2011, 32(2): 109-118.
7. Hassani B., Hinton E. A review of homogenization and topology optimization I—homogenization theory for media with periodic structure [J]. Computers and Structures, 1998, 69(6): 707-717.
8. Bendsøe M. P., Sigmund O. Material interpolation schemes in topology optimization [J]. Archive of Applied Mechanics, 1999, 69(9-10): 635-654.
9. Guedes J. M., Kikuchi N. Preprocessing and postprocessing for materials based on the homogenization method with adaptive finite element methods [J]. Computer Methods in Applied Mechanics and Engineering, 1990, 83(2): 143-198.
10. Haftka R. T. Techniques for thermal sensitivity analysis [J]. International Journal for Numerical Methods in Engineering, 1981, 17(1): 71-80.

11. Huet C. Application of variational concepts to size effects in elastic heterogeneous bodies [J]. Journal of the Mechanics and Physics of Solids, 1990, 38(6): 813-841.
12. Balendran B., Nemat-Nasser S. Bounds on elastic moduli of composites [J]. Journal of the Mechanics and Physics of Solids, 1995, 43(11): 1825-1853.
13. Hou T., Wu X. H. A multiscale finite element method for elliptic problems in composite materials and porous media [J]. Journal of Computational Physics, 1997, 134(1): 169-189.
14. Yi S. A., Xu L., Cheng G. D., et al. FEM formulation of homogenization method for effective properties of periodic heterogeneous beam and size effect of basic cell in thickness direction [J]. Computers and Structures, 2015, 156: 1-11.
15. Cheng G. D., Cai Y. W., Xu L. Novel implementation of homogenization method to predict effective properties of periodic materials [J]. Acta Mechanica Sinica, 2013, 29(4): 550-556.
16. Sigmund O. Materials with prescribed constitutive parameters: an inverse homogenization problem [J]. International Journal of Solids and Structures, 1994, 31(17): 2313-2329.
17. Bendsøe M. P. Optimal shape design as a material distribution problem [J]. Structural Optimization, 1989, 1(4): 193-202.
18. Hohe J., Becker W. Effective stress-strain relations for two-dimensional cellular sandwich cores: Homogenization, material models, and properties [J]. Applied Mechanics Reviews, 2002, 55(1): 61-87.
19. Efendiev Y., Hou T. Multiscale finite element methods for porous media flows and their applications [J]. Applied Numerical Mathematics, 2007, 57(5–7): 577-596.
20. Hassani B., Hinton E. A review of homogenization and topology optimization III—topology optimization using optimality criteria [J]. Computers and Structures, 1998, 69(6): 739-756.
21. Rozvany G. I. N., Zhou M., Birker T. Generalized shape optimization without homogenization [J]. Structural Optimization, 1992, 4(3-4): 250-252.
22. Brittain S. T., Sugimura Y., Schueller O. J. A., et al. Fabrication and mechanical performance of a mesoscale space-filling truss system [J]. Journal of Microelectromechanical Systems, 2001, 10(1): 113-120.
23. Lu T. J., liu T., Deng Z. C. Multifuctional design of cellular metals: a review [J]. Mechanics in Engineering, 2008, 30(1): 1-9.

24. Rodrigues H., Guedes J. M., Bendsoe M. P. Hierarchical optimization of material and structure [J]. Structural and Multidisciplinary Optimization, 2002, 24(1): 1-10.
25. Coelho P. G., Fernandes P. R. , Guedes J. M., et al. A hierarchical model for concurrent material and topology optimisation of three-dimensional structures [J]. Structural and Multidisciplinary Optimization, 2008, 35(2): 107-115.
26. Liu L., Yan J., Cheng G. D. Optimum structure with homogeneous optimum truss-like material [J]. Computers and Structures, 2008, 86(13): 1417-1425.
27. Niu B., Yan J., Cheng G. D. Optimum structure with homogeneous optimum cellular material for maximum fundamental frequency [J]. Structural and Multidisciplinary Optimization, 2009, 39(2): 115-132.
28. Pizzolato A., Sharma A., Maute K., et al. Multi-scale topology optimization of multi-material structures with controllable geometric complexity — applications to heat transfer problems [J]. Computer Methods in Applied Mechanics and Engineering, 2019, 357: 112552.
29. Zhao J. , Yoon H. , Youn B. D. Concurrent topology optimization with uniform microstructure for minimizing dynamic response in the time domain [J]. Computers and Structures, 2019, 221: 98-117.
30. Liang X., Du J. B. Concurrent multi-scale and multi-material topological optimization of vibro-acoustic structures [J]. Computer Methods in Applied Mechanics and Engineering, 2019, 349: 117-148.
31. Gao J., Luo Z., Li H., et al. Dynamic multiscale topology optimization for multi-regional micro-structured cellular composites [J]. Composite Structures, 2019, 211: 401-417.
32. Fu J., Li H., Gao L., et al. Design of shell-infill structures by a multiscale level set topology optimization method [J]. Computers and Structures, 2018, 212: 162-172.
33. Deng J. D., Chen W. Concurrent topology optimization of multiscale structures with multiple porous materials under random field loading uncertainty [J]. Structural and Multidisciplinary Optimization, 2017, 56(1): 1-9.
34. Cai J. H., Wang C. J., Fu Z. Robust concurrent topology optimization of multiscale structure under single or multiple uncertain load cases [J]. International Journal for Numerical Methods in Engineering, 2020, 121(7): 1456-1483.

35. Xia Q., Wang M. Y., Shi T. L. A method for shape and topology optimization of truss-like structure [J]. Structural and Multidisciplinary Optimization, 2013, 47(5): 687-697.
36. Zhao J. Q., Zhang M., Zhu Y., et al. Concurrent optimization of additive manufacturing fabricated lattice structures for natural frequencies [J]. International Journal of Mechanical Sciences, 2019, 163: 105153.
37. Zhang W. H., Sun S. P. Scale-related topology optimization of cellular materials and structures [J]. International Journal for Numerical Methods in Engineering, 2006, 68(9): 993-1011.
38. Dai G. M., Zhang W. H. Size effects of basic cell in static analysis of sandwich beams [J]. International Journal of Solids and Structures, 2008, 45(9): 2512-2533.
39. Yan J., Cheng G. D., Liu S. T., et al. Comparison of prediction on effective elastic property and shape optimization of truss material with periodic microstructure [J]. International Journal of Mechanical Sciences, 2006, 48(4): 400-413.
40. Tekoglu C., Onck P. R. Size effects in the mechanical behavior of cellular materials [J]. Journal of Materials Science, 2005, 40(22): 5911-5917.
41. Xie Y. M., Zuo Z. H., Huang X. D., et al. Convergence of topological patterns of optimal periodic structures under multiple scales [J]. Structural and Multidisciplinary Optimization, 2012, 46(1): 41-50.
42. Lipperman F., Fuchs M. B., Ryvkin M. Stress localization and strength optimization of frame material with periodic microstructure [J]. Computer Methods in Applied Mechanics and Engineering, 2008, 197(45): 4016-4026.
43. Cheng K. T., Olhoff N. An investigation concerning optimal design of solid elastic plates [J]. International Journal of Solids and Structures, 1981, 17(3): 305-323.
44. Allaire G., Bonnetier E., Francfort G., et al. Shape optimization by the homogenization method [J]. Numerische Mathematik, 1997, 76(1): 27-68.
45. Sigmund O. A 99 line topology optimization code written in Matlab [J]. Structural and Multidisciplinary Optimization, 2001, 21(2): 120-127.
46. Rodrigues H., Fernandes P. A material based model for topology optimization of thermoelastic structures [J]. International Journal for Numerical Methods in Engineering, 1995, 38(12): 1951-1965.

47. Pedersen P., Pedersen N. L. Strength optimized designs of thermoelastic structures [J]. Structural and Multidisciplinary Optimization, 2010, 42(5): 681-691.
48. Li Q., Steven G. P., Xie Y. M. Displacement minimization of thermoelastic structures by evolutionary thickness design [J]. Computer Methods in Applied Mechanics and Engineering, 1999, 179(3–4): 361-378.
49. Deaton J. D., Grandhi R. V. Stiffening of restrained thermal structures via topology optimization [J]. Structural and Multidisciplinary Optimization, 2013, 48(4): 731-745.
50. Wang B., Cheng G. D. Design of cellular structures for optimum efficiency of heat dissipation [J]. Structural and Multidisciplinary Optimization, 2005, 30(6): 447-458.
51. Gao T., Zhang W. H. Topology optimization involving thermo-elastic stress loads [J]. Structural and Multidisciplinary Optimization, 2010, 42(5): 725-738.
52. Xia Q., Wang M. Y. Topology optimization of thermoelastic structures using level set method [J]. Computational Mechanics, 2008, 42(6): 837-857.
53. Zhang W. H., Yang J. G., Xu Y. J., et al. Topology optimization of thermoelastic structures: mean compliance minimization or elastic strain energy minimization [J]. Structural and Multidisciplinary Optimization, 2014, 49(3): 417-429.
54. Wang B., Yan J., Cheng G. D. Optimal structure design with low thermal directional expansion and high stiffness [J]. Engineering Optimization, 2011, 43(6): 581-595.
55. Gao T., Xu P. L., Zhang W. H. Topology optimization of thermo-elastic structures with multiple materials under mass constraint [J]. Computers and Structures, 2016, 173: 150-160.
56. Wu C., Fang J. G., Li Q. Multi-material topology optimization for thermal buckling criteria [J]. Computer Methods in Applied Mechanics and Engineering, 2018, 346: 1136-1155.
57. Zhu J. H., Li Y., Wang F. W., et al. Shape preserving design of thermo-elastic structures considering geometrical nonlinearity [J]. Structural and Multidisciplinary Optimization, 2020, 61(5): 1787-1804.
58. Liu T., Deng Z. C., Lu T. J. Design optimization of truss-cored sandwiches with homogenization [J]. International Journal of Solids and Structures, 2006, 43(25–26): 7891-7918.

59. Taylor C. M., Smith C. W., Miller W., et al. The effects of hierarchy on the in-plane elastic properties of honeycombs [J]. International Journal of Solids and Structures, 2011, 48(9): 1330-1339.
60. Deng J. D., Yan J., Cheng G. D. Multi-objective concurrent topology optimization of thermoelastic structures composed of homogeneous porous material [J]. Structural and Multidisciplinary Optimization, 2013, 47(4): 583-597.
61. Sigmund O., Torquato S. Design of materials with extreme thermal expansion using a three-phase topology optimization method [J]. Proceedings of SPIE - The International Society for Optical Engineering, 1997, 45(6): 1037-1067.
62. Radman A., Huang X., Xie Y. M. Topological design of microstructures of multi-phase materials for maximum stiffness or thermal conductivity [J]. Computational Materials Science, 2014, 91(91): 266-273.
63. Pelanconi M., Barbato M., Zavattoni S., et al. Thermal design, optimization and additive manufacturing of ceramic regular structures to maximize the radiative heat transfer [J]. Materials and Design, 2019, 163: 107539.
64. Niknam H., Akbarzadeh A. H. Thermo-mechanical bending of architected functionally graded cellular beams [J]. Composites Part B-Engineering, 2019, 174: 107060.
65. Wu T., Tovar A. Multiscale, thermomechanical topology optimization of self-supporting cellular structures for porous injection molds [J]. Rapid Prototyping Journal, 2019, 25(9): 1482-1492.
66. Xu B., Jiang J. S., Xie Y. M. Concurrent design of composite macrostructure and multi-phase material microstructure for minimum dynamic compliance [J]. Composite Structures, 2015, 128: 221-233.
67. Xu B., Huang X., Zhou S. W., et al. Concurrent topological design of composite thermoelastic macrostructure and microstructure with multi-phase material for maximum stiffness [J]. Composite Structures, 2016, 150: 84-102.
68. Das S., Sutradhar A. Multi-physics topology optimization of functionally graded controllable porous structures: application to heat dissipating problems [J]. Materials and Design, 2020, 193.
69. Valentin J., Huebner D., Stingl M., et al. Gradient-based two-scale topology optimization with B-splines on sparse grids [J]. Siam Journal on Scientific Computing, 2020, 42(4): B1092-B1114.

70. Shanyi D. U. Advanced composite materials and aerospace engineering [J]. Acta Materiae Compositae Sinica, 2007, 24(1): 1-12.
71. Reddy J. N., Khdeir A. Buckling and vibration of laminated composite plates using various plate theories [J]. AIAA Journal, 1989, 27(12): 1808-1817.
72. Stegmann J., Lund E. Discrete material optimization of general composite shell structures [J]. International Journal for Numerical Methods in Engineering, 2005, 62(14): 2009-2027.
73. Liu S. T., Hou Y. P., Sun X. N., et al. A two-step optimization scheme for maximum stiffness design of laminated plates based on lamination parameters [J]. Composite Structures, 2012, 94(12): 3529-3537.
74. Ferreira R. T. L., Rodrigues H. C., Guedes J. M., et al. Hierarchical optimization of laminated fiber reinforced composites [J]. Composite Structures, 2014, 107: 246-259.
75. Callahan K. J., Weeks G. E. Optimum design of composite laminates using genetic algorithms [J]. Composites Engineering, 1992, 2(3): 149-160.
76. Nagendra S., Jestin D., Gürdal Z., et al. Improved genetic algorithm for the design of stiffened composite panels [J]. Computers and Structures, 1996, 58(3): 543-555.
77. Park J. H., Hwang J. H., Lee C. S., et al. Stacking sequence design of composite laminates for maximum strength using genetic algorithms [J]. Composite Structures, 2001, 52(2): 217-231.
78. Graesser D. L., Zabinsky Z. B., Tuttle M. E., et al. Designing laminated composites using random search techniques [J]. Composite Structures, 1991, 18(4): 311-325.
79. Ghiasi H., Pasini D., Lessard L. Optimum stacking sequence design of composite materials Part I: constant stiffness design [J]. Composite Structures, 2009, 90(1): 1-11.
80. Ghiasi H., Fayazbakhsh K., Pasini D., et al. Optimum stacking sequence design of composite materials Part II: variable stiffness design [J]. Composite Structures, 2010, 93(1): 1-13.
81. Bakis C. E., Bank L. C., Brown V. L. , et al. Fiber-reinforced polymer composites for construction--state-of-the-art review [J]. Journal of Composites for Construction, 2002, 6(2): 73-87.
82. Ganguli R. J. Optimal design of composite structures: a historical review [J]. Journal of the Indian Institute of Science, 2013, 93: 557-570.

83. Nikbakt S., Kamarian S., Shakeri M. A review on optimization of composite structures part I: laminated composites [J]. Composite Structures, 2018, 195: 158-185.
84. Xu Y. J., Zhu J. H., Wu Z., et al. A review on the design of laminated composite structures: constant and variable stiffness design and topology optimization [J]. Advanced Composites and Hybrid Materials, 2018, 1(3): 460-477.
85. Lund E., Stegmann J. On structural optimization of composite shell structures using a discrete constitutive parametrization [J]. Wind Energy, 2005, 8(1): 109-124.
86. Sørensen S. N., Sørensen R., Lund E. DMTO – a method for discrete material and thickness optimization of laminated composite structures [J]. Structural and Multidisciplinary Optimization, 2014, 50(1): 25-47.
87. Wu C., Gao Y. K., Fang J. G., et al. Simultaneous discrete topology optimization of ply orientation and thickness for carbon fiber reinforced plastic-laminated structures [J]. Journal of Mechanical Design, 2019, 141(4): 044501.
88. Niu B., Olhoff N., Lund E., et al. Discrete material optimization of vibrating laminated composite plates for minimum sound radiation [J]. International Journal of Solids and Structures, 2010, 47(16): 2097-2114.
89. Kiyono C. Y., Silva E. C. N., Reddy J. N. A novel fiber optimization method based on normal distribution function with continuously varying fiber path [J]. Composite Structures, 2017, 160: 503-515.
90. Hao P., Feng S. J., Zhang K., et al. Adaptive gradient-enhanced kriging model for variable-stiffness composite panels using Isogeometric analysis [J]. Structural and Multidisciplinary Optimization, 2018, 58(1): 1-16.
91. Akhavan H., Ribeiro P. Aeroelasticity of composite plates with curvilinear fibres in supersonic flow [J]. Composite Structures, 2018, 194: 335-344.
92. Hao P., Liu D., Wang Y., et al. Design of manufacturable fiber path for variable-stiffness panels based on lamination parameters [J]. Composite Structures, 2019, 219: 158-169.
93. Shafighfard T., Demir E., Yildiz M. Design of fiber-reinforced variable-stiffness composites for different open-hole geometries with fiber continuity and curvature constraints [J]. Composite Structures, 2019, 226: 111280.

94. Serhat G., Bediz B., Basdogan I. Unifying lamination parameters with spectral-tchebychev method for variable-stiffness composite plate design [J]. Composite Structures, 2020, 242: 112183.
95. Bruyneel M. SFP—a new parameterization based on shape functions for optimal material selection: application to conventional composite plies [J]. Structural and Multidisciplinary Optimization, 2011, 43(1): 17-27.
96. Gao T., Zhang W. H., Pierre D. A bi-value coding parameterization scheme for the discrete optimal orientation design of the composite laminate [J]. Numerical method in engineering, 2012, 91(1): 98-114.
97. Hvejsel C. F., Lund E. Material interpolation schemes for unified topology and multi-material optimization [J]. Structural and Multidisciplinary Optimization, 2011, 43(6): 811-825.
98. Gao T., Zhang W. H. A mass constraint formulation for structural topology optimization with multiphase materials [J]. International Journal for Numerical Methods in Engineering, 2011, 88(8): 774-796.
99. Hao P., Liu X. X., Wang Y., et al. Collaborative design of fiber path and shape for complex composite shells based on isogeometric analysis [J]. Computer Methods in Applied Mechanics and Engineering, 2019, 354: 181-212.
100. Hao P., Yuan X., Liu C., et al. An integrated framework of exact modeling, isogeometric analysis and optimization for variable-stiffness composite panels [J]. Computer Methods in Applied Mechanics and Engineering, 2018, 339: 205-238.
101. Duan Z., Yan J., Zhao G. Integrated optimization of the material and structure of composites based on the heaviside penalization of discrete material model [J]. Structural and Multidisciplinary Optimization, 2015, 51(3): 721-732.
102. Guest J. K., Prévost J. H., Belytschko T. Achieving minimum length scale in topology optimization using nodal design variables and projection functions [J]. International Journal for Numerical Methods in Engineering, 2004, 61(2): 238-254.
103. Guest J. K. Imposing maximum length scale in topology optimization [J]. Structural and Multidisciplinary Optimization, 2009, 37(5): 463-473.
104. Sigmund O. Morphology-based black and white filters for topology optimization [J]. Structural and Multidisciplinary Optimization, 2007, 33(4-5): 401-424.

105. Yan J., Hu W. B., Duan Z. Y. Structure/material concurrent optimization of lattice materials based on extended multiscale finite element method [J]. International Journal for Multiscale Computational Engineering, 2015, 13(1): 73-90.
106. Haber R. B., Jog C. S., Bendsøe M. P. A new approach to variable-topology shape design using a constraint on perimeter [J]. Structural Optimization, 1996, 11(1): 1-12.
107. Sigmund O. Design of Material Structures Using Topology Optimization [M]. Frontiers, 1994.
108. Petersson J., Sigmund O. Slope constrained topology optimization [J]. International Journal for Numerical Methods in Engineering, 1998, 41: 1417-1434.
109. Zhang W. H, Duysinx P. Dual approach using a variant perimeter constraint and efficient sub-iteration scheme for topology optimization [J]. Computers and Structures, 2003, 81(22): 2173-2181.
110. Boggs P. T., Tolle J. W. Sequential quadratic programming [J]. Acta Numerica, 1995, 4: 1-51.
111. Liu S. T., Cheng G. D., Gu Y., et al. Mapping method for sensitivity analysis of composite material property [J]. Structural and Multidisciplinary Optimization, 2002, 24(24): 212-217.
112. Wang A. J, Mcdowell D. L. In-plane stiffness and yield strength of periodic metal honeycombs [J]. Journal of Engineering Materials and Technology, 2004, 126(2): 137-156.
113. Guo X., Zhang W. S., Zhong W. L. Stress-related topology optimization of continuum structures involving multi-phase materials [J]. Computer Methods in Applied Mechanics and Engineering, 2014, 268: 632–655.
114. Yan J., Cheng G. D., Liu L., et al. Concurrent material and structural optimization of hollow plate with truss-like material [J]. Structural and Multidisciplinary Optimization, 2008, 35(2): 153-163.
115. Yan J., Hu W. B., Wang Z. H., et al. Size effect of lattice material and minimum weight design [J]. Acta Mechanica Sinica, 2014, 30(02): 191-197.
116. Pecullan S., Gibiansky L. V., Torquato S. Scale effects on the elastic behavior of periodic and hierarchical two-dimensional composites [J]. Journal of the Mechanics and Physics of Solids, 1999, 47(7): 1509-1542.
117. Fleck N. A., Hutchinson J. W. Strain gradient plasticity [J]. Advances in Applied Mechanics, 1997, 33: 295-361.

118. Adachi T., Tomita Y., Tanaka M. Computational simulation of deformation behavior of 2D-lattice continuum [J]. International Journal of Mechanical Sciences, 1998, 40(9): 857-866.
119. Eringen A. C. Microcontinuum field theories. I. Foundations and solids [M]. 1999.
120. Chen J. Y., Huang Y., Ortiz M. Fracture analysis of cellular materials: a strain gradient model [J]. Journal of the Mechanics and Physics of Solids, 1998, 46(5): 789–828.
121. Lipton R. Bounds on the distribution of extreme values for the stress in composite materials [J]. Journal of the Mechanics and Physics of Solids, 2004, 52(5): 1053-1069.
122. Lipton R., Stuebner M. Inverse homogenization and design of microstructure for pointwise stress control [J]. Quarterly Journal of Mechanics and Applied Mathematics, 2006, 59(1): 139-161(123).
123. Kumar R. S., McDowell D. L. Generalized continuum modeling of 2-D periodic cellular solids [J]. International Journal of Solids and Structures, 2004, 41(26): 7399-7422.
124. Sachio N., Benedict R., Lakes R. Finite element method for orthotropic micropolar elasticity [J]. International Journal of Engineering Science, 1984, 22(3): 319-330.
125. Gibson L. J., Ashby M. F. Cellular solids: structure and properties. cambridge [M]. Cambridge University Press, 1997.
126. Cochran J. K., Lee K. J., McDowell D., et al. Multifunctional metallic honeycombs by thermal chemical processing [C]. Processsing and Properties of Lightweight Cellular Metals and Structures. Warrendale. 2002.
127. Cook R. D., Malkus D. S., Plesha M. E. Concepts and applications of finite element analysis. [M]. New York: Wiley, 2002.
128. Cheng G. D. Topology optimization for elasticity [M]. Dalian: Dalian University of Technology Press, 1985.
129. Liu T., Deng Z. C., Lu T. J. Minimum weights of pressurized hollow sandwich cylinders with ultralight cellular cores [J]. International Journal of Solids and Structures, 2007, 44(10): 3231-3266.
130. Xu S. L., Cai Y. W., Cheng G. D. Volume preserving nonlinear density filter based on heaviside functions [J]. Structural and Multidisciplinary Optimization, 2010, 41(4): 495-505.

131. Gournay F. D. Velocity extension for the level-set method and multiple eigenvalues in shape optimization [J]. Siam Journal on Control and Optimization, 2005, 45(1): 343-367.
132. Du J. B., Olhoff N. Topological design of freely vibrating continuum structures for maximum values of simple and multiple eigenfrequencies and frequency gaps [J]. Structural and Multidisciplinary Optimization, 2007, 34(2): 91-110.
133. Ma Z. D., Kikuchi N., Cheng H. C. Topological design for vibrating structures [J]. Computer Methods in Applied Mechanics and Engineering, 1995, 121(s 1–4): 259-280.
134. Gomes F. A. M., Senne T. A. An SLP algorithm and its application to topology optimization [J]. Computational and Applied Mathematics, 2011, 30(1): 53-89.
135. Haug E., Rousselet B. Design sensitivity analysis in structural mechanics.II. eigenvalue bariations [J]. Journal of Structural Mechanics, 1980, 8(2): 161-186.
136. Seyranian A. P., Lund E., Olhoff N. Multiple eigenvalues in structural optimization problems [J]. Structural Optimization, 1994, 8(4): 207-227.
137. Cox S. J. The generalized gradient at a multiple eigenvalue [J]. Journal of Functional Analysis, 1995, 133(1): 30-40.
138. Pedersen N. L. Maximization of eigenvalues using topology optimization [J]. Structural and Multidisciplinary Optimization, 2000, 20(1): 2-11.
139. Allaire G., Aubry S., Jouve F. Eigenfrequency optimization in optimal design [J]. Computer Methods in Applied Mechanics and Engineering, 2001, 190(28): 3565-3579.
140. Cheng G. D., Wang B. Constraint continuity analysis approach to structural topology optimization with frequency objective [C]. Kwak BM et al (eds) Proceeding of the 7th world congress of structural and multidisciplinary optimization. Seoul, Korea. 2007.
141. Bendsøe M. P., Sigmund O. Topology optimization - theory, methods, and applications [M]. Berlin Heidelberg: Springer Verlag, 2003.
142. Artola M., Duvaut G. Homogénéisation d'une plaque renforcée [J]. Comptes Rendus Hebdomadaires des Séances de l'Académie des Sciences Série A Sciences Mathématiques, 1977, 12: A707-A710.
143. Liu S. T., Zhang J. H. Homogenization-based method for bending analysis of perforated plate [J]. Journal of Solid Mechanics, 1998, 11(2): 172-179.

144. Yan J., Guo X., Cheng G. D. Multi-scale concurrent material and structural design under mechanical and thermal loads [J]. Computational Mechanics, 2016, 57(3): 437-446.
145. Yan J., Cheng G. D., Liu L. A uniform optimum material based model for concurrent optimization of thermoelastic structures and materials [J]. International Journal for Simulation and Multidisciplinary Design Optimization, 2008, 2(4): 259-266.
146. Liu S. T., Cheng G. D. Homogenization-based method for predicting thermal expansion coefficients of composite materials [J]. Journal of Dalian University of Technology, 1995, 35.
147. Hashin Z. Statistical cumulative damage theory for fatigue Life prediction [J]. Journal of Applied Mechanics, 1983, 50(3): 571-579.
148. Svanberg K. A class of globally convergent optimization methods based on conservative convex separable approximations [J]. Siam Journal on Optimization, 2009, 12(2): 555-573.
149. Patzák B., Bittnar Z. Design of object oriented finite element code [J]. Advances in Engineering Software, 2001, 32(10): 759-767.
150. Kruijf N. D., Zhou S. W., Li Q., et al. Topological design of structures and composite materials with multiobjectives [J]. International Journal of Solids and Structures, 2007, 44(22): 7092-7109.
151. Yan J., Duan Z. Y., Lund E., et al. Concurrent multi-scale design optimization of composite frame structures using the Heaviside penalization of discrete material model [J]. Acta Mechanica Sinica, 2016, 32(3): 430-441.
152. Yan J., Duan Z. Y., Lund E., et al. Concurrent multi-scale design optimization of composite frames with manufacturing constraints [J]. Structural and Multidisciplinary Optimization, 2017, 56(3): 519-533.
153. Duan Z. Y., Yan J., Lee I., et al. Integrated design optimization of composite frames and materials for maximum fundamental frequency with continuous fiber winding angles [J]. Acta Mechanica Sinica, 2018, 34(6): 1084-1094.
154. Duan Z. Y., Yan J., Lee I., et al. Discrete material selection and structural topology optimization of composite frames for maximum fundamental frequency with manufacturing constraints [J]. Structural and Multidisciplinary Optimization, 2019, 60(5): 1741-1758.
155. Bailie J. A., Ley R. P., Pasricha A., A summary and review of composite laminate design guidelines [R]. NAS1-19347: Northrop Grumman-Military Aircraft Systems Division, 1997.

156. Deaton J. D., Grandhi R. V. A survey of structural and multidisciplinary continuum topology optimization: post 2000 [J]. Structural and Multidisciplinary Optimization, 2014, 49(1): 1-38.
157. Guest J. K. Eliminating beta-continuation from Heaviside projection and density filter algorithms [J]. Structural and Multidisciplinary Optimization, 2011, 44(4): 443-453.
158. Blasques J. P., Lazarov B. BECAS - a beam cross-section analysistool for anisotropic and inhomogeneous sections of arbitrary geometry [M]. Risø National Lab, Technical University of Denmark: Technical report, 2011.
159. Baker A. A., Dutton S. E., Kelly D. W. Composite materials for aircraft structures / 2nd ed [M]. American Institute of Aeronautics and Astronautics, 2004.
160. Duan Z. Y., Yan J., Niu B., et al. Structural optimization design of composite materials based on improved discrete material interpolation format [J]. Journal of Aviation, 2012, (12): 2221-2229.
161. Stegmann J., Lund E. Nonlinear topology optimization of layered shell structures [J]. Structural and Multidisciplinary Optimization, 2005, 29(5): 349-360.
162. Blasques J. P., Stolpe M. Maximum stiffness and minimum weight optimization of laminated composite beams using continuous fiber angles [J]. Structural and Multidisciplinary Optimization, 2010, 43(4): 573-588.
163. Lund E. Finite element based design sensitivity analysis and optimization [M]. Aalborg University, 1994.
164. Cheng G., Olhoff N. Rigid body motion test against error in semi-analytical sensitivity analysis [J]. Computers and Structures, 1993, 46(3): 515-527.
165. Cheng G. D., Liu Y. W. A new computation scheme for sensitivity analysis [J]. Engineering Optimization, 1987, 12(3): 219-234.
166. Dorn W. S., Gomory R. E., Greenberg H. Automatic design of optimal structures [J]. Journal de mecanique, 1964, (3): 25-52.
167. Mallick P. K. Fiber-reinforced composites: materials, manufacturing, and design [M]. 2007.
168. Bruyneel M., Beghin C., Craveur G., et al. Stacking sequence optimization for constant stiffness laminates based on a continuous optimization approach [J]. Structural and Multidisciplinary Optimization, 2012, 46(6): 783-794.

169. Seresta O., Gürdal Z., Adams D. B., et al. Optimal design of composite wing structures with blended laminates [J]. Composites Part B: Engineering, 2007, 38(4): 469-480.
170. Kassapoglou C. Design and analysis of composite structures: with applications to aerospace structures, 2nd edn [M]. Sons: John Wiley & Sons Ltd, 2013.
171. Irisarri F. X., Lasseigne A., Leroy F. H., et al. Optimal design of laminated composite structures with ply drops using stacking sequence tables [J]. Composite Structures, 2014, 107(1): 559-569.
172. Lund E., Sørensen S. N. Topology and thickness optimization of laminated composites including manufacturing constraints [M]. Springer-Verlag New York, Inc., 2013.
173. Blasques J. P. Multi-material topology optimization of laminated composite beams with eigenfrequency constraints [J]. Composite Structures, 2014, 111(1): 45-55.
174. Blasques, Amaral J. P. A. User's manual for BECAS : a cross section analysis tool for anisotropic and inhomogeneous beam sections of arbitrary geometry. 2012.
175. Blasques J. P., Stolpe M. Multi-material topology optimization of laminated composite beam cross sections [J]. Composite Structures, 2012, 94(11): 3278-3289.

Index

Printed in the United States
by Baker & Taylor Publisher Services